T0256077

Science in the Age of Computer Simulation

Science in the Age of Computer Simulation

ERIC B. WINSBERG

The University of Chicago Press   Chicago and London

ERIC B. WINSBERG is associate professor of philosophy at the
University of South Florida.

The University of Chicago Press, Chicago 60637
The University of Chicago Press, Ltd., London
© 2010 by The University of Chicago
All rights reserved. Published 2010
Printed in the United States of America

19 18 17 16 15 14 13 12 11 10   1 2 3 4 5

ISBN-13: 978-0-226-90202-9 (cloth)
ISBN-13: 978-0-226-90204-3 (paper)
ISBN-10: 0-226-90202-1 (cloth)
ISBN-10: 0-226-90204-8 (paper)

Library of Congress Cataloging-in-Publication Data
Winsberg, Eric B.
    Science in the age of computer simulation / Eric B. Winsberg.
        p. cm.
    Includes bibliographical references and index.
    ISBN-13: 978-0-226-90202-9 (cloth : alk. paper)
    ISBN-13: 978-0-226-90204-3 (pbk. : alk. paper)
    ISBN-10: 0-226-90202-1 (cloth : alk. paper)
    ISBN-10: 0-226-90204-8 (pbk. : alk. paper) 1. Science—Data
processing—Philosophy. 2. Knowledge, Theory of. 3. Science—
Experiments—Computer simulation. 4. Science—Methodology.
5. Science—Philosophy. I. Title.
    Q175.32 .K46W56 2010
    507.2—dc22   2010013703

⊗ The paper used in this publication meets the minimum requirements
of the American National Standard for Information Sciences—
Permanence of Paper for Printed Library Materials, ANSI Z39.48–1992.

This book is dedicated to the memory of

SUZANNE G. A. WINSBERG

AND

FRED WINSBERG

Contents

Acknowledgments

I want to begin by thanking those people who encouraged and mentored my early work on the subject of this book: Stephen Kellert, Michael Friedman, and Frederick Suppe at Indiana University. My first work on material that is incorporated in this book was supported by a postdoctoral fellowship in the History and Philosophy of Science program at Northwestern University, where I had the privilege to work with Arthur Fine. This was a wonderful, intellectually stimulating environment in which much of the philosophical framework of this book emerged. For this, I have not only Arthur Fine to thank, but also Mathias Frisch, with whom I shared then, and have since, numerous conversations and debates that have shaped many of my philosophical intuitions over the years.

Work on this manuscript began in earnest while I was a fellow at the Zentrum für interdisziplinäre Forschung (ZiF) at the University of Bielefeld in the research group Science in the Context of Application. I am extremely grateful for their financial support. I am also grateful for the hospitality, generosity, and intellectual camaraderie of the organizers, Martin Carrier and Alfred Nordmann. I also want to thank Torsten Wilholt, Justin Biddle, Johannes Lehnard, Felicitas Krämer, and too many other people to mention, mostly postdocs and graduate students, for making my time in Bielefeld fun and intellectually productive. My work on the manuscript continued while I was a fellow of the Institute of Advanced Study (IAS) at Durham University. While working at the IAS, I was hosted by University College, Durham. I am grateful to both of these institutions for supporting my work. I am also grateful to all the other IAS fellows who shared my time there, especially the statisticians: Peter Challenor and John Hasslet, who shared with me some of their expertise on the estimation of uncertainties associated with climate models. And, of course, much of the work on this book was completed while I have been a professor in the philosophy department at the University of South Florida. I want to thank all of my colleagues in the department for their support and conversations, especially Daniel Weiskopf. I also want to thank Wendy Parker, with whom I have been exchanging manuscript drafts via e-mail for years. Finally, I want to thank three senior philosophers of science with whom I have never shared residence, but who have supported me and my work virtually since I left graduate school: Margaret Morrison, Paul Teller, and Ronald Giere.

Some of this book is based on work that has appeared elsewhere. This includes "Sanctioning Models: The Epistemology of Simulation," which appeared in *Science in Context*; "Simulated Experiments: Methodology for a Virtual World" and "Simulations, Models and Theories: Complex Physical Systems and Their Representations," both of which appeared in *Philosophy of Science*; "A Tale of Two Methods," which appeared in *Synthese*;[1] "Handshaking Your Way to the Top: Simulation at the Nanoscale," which appeared in *Philosophy of Science*; "A Function for Fictions: Expanding the Scope of Science," which appeared in *Fictions in Science: Philosophical Essays on Modeling and Idealization,* edited by Mauricio Suarez and published by Routledge; and "Models of Success vs. the Success of Models: Reliability without Truth," which was published in *Synthese*.[2] Chapter 6 contains material from two papers that were coauthored. One is the not-yet-published "Holism and Entrenchment in Climate Models," which I wrote with Johannes Lenhard. The other is "Value Judgments and the Estimation of Uncertainty in Climate Modeling," which I wrote with Justin Biddle, and which appeared in *New Waves in the Philosophy of Science*, edited by P. D. Magnus and J. Busch, published by Palgrave Mac-Millan. I want to thank Johannes and Justin both very much for allowing me to borrow freely from those two papers for this book.

I also want to thank all of the people who were directly involved with making this book come into being at the University of Chicago Press. Three different reviewers made helpful comments on earlier drafts of the

1. With kind permission from Springer Science+Business Media: Eric Winsberg, "A Tale of Two Methods," *Synthese* 169 (2009): 575–92. © Springer Science+Business Media B. V. 2008.

2. With kind permission from Springer Science+Business Media: Eric Winsberg, "Models of Success vs. the Success of Models: Reliability without Truth," *Synthese* 152 (2006): 1–19. © Springer 2006.

manuscript. Two of them were anonymous, but the one who gave the most detailed and helpful suggestions was not: Robert Rosner, a physicist and the director of the Argonne National Laboratory. All three of them have my thanks for making this a better book. Thanks especially to Karen Darling, my editor, who has been extremely helpful and supportive all the way through the process, to my copy editor, Nick Murray, and to the rest of the staff at the University of Chicago Press.

Most important, I want to thank Laura, who has tirelessly supported me in this work all the way from the beginning.

ONE

Introduction

Major developments in the history of the philosophy of science have always been driven by major developments in the sciences. The most famous examples, of course, are the revolutionary changes in physics at the beginning of the twentieth century that inspired the logical positivists of the Vienna Circle. But there are many others. Kant's conception of synthetic *a priori* knowledge was originally intended to address the new mechanics of Newton. The rise of non-Euclidean geometries in the nineteenth century led to Helmholtz's revised formulation of transcendentalism, as well as, more famously, to Poincaré's defense of conventionalism. The rise of atomic theory in the nineteenth century and the ensuing skepticism about the genuine existence of atoms, to raise one final example, played a large role in igniting and fueling debates about scientific realism that continue to rage today.

Over the last fifty years, however, there has been a revolutionary development affecting almost all of the sciences that, at least until very recently, has been largely ignored by philosophers of science. The development I am speaking of is the astonishing growth, in almost all of the sciences, of the use of the digital computer to study phenomena of great complexity—the rise of computer simulations. More and more scientific "experiments" are, to use the vernacular of the day, being carried out "in silico."

It is certainly true that, historically, most of the famous scientific developments that have had an impact on the philosophy of science have involved revolutionary changes at the level of fundamental theory. It is also true that the use of computer simulation to study complex phenomena usually occurs against a backdrop of well-established basic theory, rather than in the process of altering, let alone revolutionizing, such theory. But surely there is no reason to think that it is only changes in basic theory that should be of interest to philosophers. Surely there is no reason to think that new experimental methods, new research technologies, or innovative ways of solving new sets of problems within existing theory could not have a similar impact on philosophy. It is not altogether unlikely that some of the major accomplishments in the physical sciences to come in the near future will have as much to do with modeling complex phenomena within existing theories as with developing novel fundamental theories.

That, in a nutshell, is the basic sentiment that motivates this book: that the last part of the twentieth century has been, and the twenty-first century is likely to continue to be, the age of computer simulation. This has been an era in which, at least in the physical sciences, and to a large degree elsewhere, major developments in fundamental theory have been slow to come, but there has been an avalanche of novel applications of existing theory—an avalanche aided in no small part by our increasing ability to use the digital computer to build tractable models of greater and greater complexity, using the same available theoretical resources. The book is motivated as well by the conviction that the philosophy of science should continue, as it always has in the past, to respond to the character of the science of its own era. This book, therefore, is about computer simulation and the philosophy of science; and it is as much about what philosophers of science should learn in the age of simulation as it is about what philosophy can contribute to our understanding of how the digital computer is transforming science.

Science and Its Applications

General philosophy of science concerns itself with a diverse set of issues: the nature of scientific evidence, the nature and scope of scientific theories; the relations between theories at different levels of description; the relationship between theories on the one hand and local descriptions of phenomena on the other; the role that various kinds of models play in mediating those relationships; the nature of scientific explanation; and the issue of scientific realism, just to name a few. Our understanding of these topics, I will argue in this book, could greatly profit from a close look at examples of scientific practice where computer simulation plays a prominent role. There are also new topics that can arise for the philosophy of science, topics that have specifically to do with simulation but are of a distinctly philosophical character. I will tackle some of these in this book: What is the relationship between computer simulation, or simulation generally, and experiment? Under what conditions should we expect a computer simulation to be reliable? How can we evaluate a simulation model when the predictions made by such a model are precisely about those phenomena for which data are sparse? What role do deliberately false assumptions play in the construction of simulation models?

Let us begin with one of the oldest topics in the philosophy of science—the nature of scientific evidence. Computer simulations are involved in the creation and justification of scientific knowledge claims, and the problem of the nature of scientific evidence in the philosophy of science is precisely the concern with saying when we do, or don't, have evidence that such claims to knowledge are justified. But simulations more often involve the *application* rather than the testing of scientific theories. And so the epistemology of simulation is a topic that is quite unfamiliar to most philosophy of science, which has traditionally concerned itself with the justification of theories, not with their application. An appropriately subtle understanding of the epistemology of simulation requires that we rethink the relationship between theories and local descriptions of phenomena.

The rethinking required dovetails nicely, moreover, with recent debates in the philosophy of science about the scope of theories. According to one side in this debate, laws and theories in science are tightly restricted with respect to the features of the world that fall under their domain. The other side maintains that fundamental theories by their nature have universal domains. Few of the simulations considered in this book have much to do with fundamental theory, and so that precise debate will not concern us directly. But there is a related question that the epistemology of simulation must confront: Does the principled scope of every theory extend as far as all of its less-than-principled applications? More concretely, when simulationists use a particular theory to guide the construction of their simulations, is it necessarily the case that their results are, properly speaking, part of the "empirical content" of those theories? This is an important question both for the general philosopher of science interested in the nature of scientific theories and, as we shall see, for anyone interested in the epistemology of simulation. To get a clearer view of these issues, we must look at some of the details of computer simulation methods.

A Brief History

Computer simulation is a method for studying complex systems that has had applications in almost every field of scientific study—from quantum chemistry to the study of traffic-flow patterns. Its history is as long as the history of the digital computer itself, and it begins in the United States during the Second World War. When the physicist John Mauchly visited the Ballistic Research Laboratory at Aberdeen and saw an army of women calculating firing tables on mechanical calculators, he suggested that the laboratory begin work on a digital computer. The Electrical Numerical Integrator and Computer (ENIAC), the first truly programmable digital computer, was born in 1945. John von Neumann took an immediate interest in it and, encouraged by fellow Hungarian-American physicist Edward Teller, he enlisted the help of Nicholas Metropolis and Stanislaw Ulam to begin work on a computational model of a thermonuclear reaction.

Their effort was typical of computer simulation techniques. They began with a mathematical model depicting the time-evolution of the system being studied in terms of equations, or rules-of-evolution, for the variables of the model. The model was constructed (as is typical in the physical sciences) from a mixture of well-established theoretical principles, some physical insight, and some clever mathematical tricks. They then transformed the model into a computable algorithm, and the evolution of the computer was said to "simulate" the evolution of the system in question.

The war ended before von Neumann's project was completed, but its eventual success persuaded Teller, von Neumann, and Enrico Fermi of the feasibility of a hydrogen bomb. It also convinced the military high brass of the practicability of electronic computation. Soon after, meteorology joined the ranks of weapons research as one of the first disciplines to make use of the computer. Von Neumann was convinced early on that hydrodynamics was very important to physics and that its development would require vast computational resources. He also became convinced that it would be strategic to enlist meteorologists, with the resources at their disposal, as allies. In 1946, he launched the Electronic Computer Project at the Institute for Advanced Study at Princeton University and chose numerical meteorology as one of its largest projects. While working on the problem of simulating simplified weather systems, meteorologist and mathematician Edward Lorentz discovered a simple model that displayed characteristics now called "sensitive dependence on initial conditions" and "strange attractors," the hallmarks of a system well described by "chaos theory," a field he helped to create.

In the last forty-some years, simulations have proliferated in the sciences, and an enormous variety of techniques have been developed. In the physical sciences, two classes of simulations predominate: continuum methods and particle methods. Continuum methods begin by describing their object of study as a medium described by fields and distributions—a continuum. The goal is then to give differential equations that relate the rates of change of the values for these fields and distributions, and to use "discretization" techniques to transform these continuous differential equations into algebraic expressions that can be calculated step-by-step by a computer. Particle methods describe their objects of study either as a collection of nuclei and electrons or as a collection of atoms and molecules—the former only if quantum methods are employed.[1]

From a certain point of view, these are methods for overcoming merely practical limitations in our abilities to solve the equations provided by our best theories—theories like fluid mechanics, quantum mechanics, and classical molecular dynamics. Why should methods for overcoming practical limitations be of interest to philosophers? Philosophers of science are accustomed to centering the attention they devote to scientific theories on a cluster of canonical issues: What are theories? How are they confirmed? How should we interpret them? They tend to think that all of the philosophically interesting action takes place around that cluster— that what matters to philosophy is the nature of theories in principle, not what we are merely limited to doing with them in practice. The practical obstacles that need to be overcome when we work with theories can strike the philosopher as mundane.

This is as good a place as any to point out a methodological presupposition that prevails in most of this book. As I said above, I believe that philosophers of science have missed an opportunity to contribute to this explosive area of modern science precisely because they have had a prejudice for being concerned with what is possible in principle rather than with what we can achieve in practice. Accordingly, I focus on the current state of practice of computer simulation, rather than on what we might think might, in principle, someday be possible. When I say in what follows, for example, that we cannot do P, I often mean that the

---

1. Particle methods often make use of "Monte Carlo" algorithms. Such methods use random sampling algorithms, where the randomness of the algorithm need not correspond to an underlying indeterminism in the system. Outside of physics, it is common to encounter "cellular automata." These are simulations that assign a discrete state to each node of a network of elements and assign rules of evolution for each node based on its local environment. Such simulations are especially common in the social sciences, where each node can be thought of as an "agent" reacting to its local environment.

prospect of doing P is presently intractable, not that proofs exist of the fact that P is impossible in principle. One might wonder whether or not it is sensible to draw philosophical conclusions from present practical difficulties—and the methods by which they are overcome—instead of focusing on what is possible or impossible in principle.

This is indeed a worry, but I do not believe it should prevent us from getting started on the difficult philosophical work of examining the present state of the art. It is true that some of the practical obstacles I discuss in what follows may someday be overcome. We may find more principled solutions to some of the problems for which we now apply less principled approaches. But I doubt this will mean that the lessons we draw from studying those less principled approaches will lose their take-home value when they are replaced. That is because I doubt we will ever reach a point where all scientific problems will have theoretically principled solutions. Scientific practice will surely evolve, but we will always be pushing the envelope of the set of problems we want to tackle with existing theory, and new practical difficulties will arise as old ones disappear. More important, philosophers should not allow the present state of flux in the computationally intensive sciences to prevent them from paying close enough attention to where most of the action has been in recent science: the unprincipled solutions of just these sorts of practical difficulties. This is a part of scientific practice that is responsible for more and more of the creation of knowledge in science and one that is ripe for philosophical attention.

And we should make no mistake about it—simulation is a process of knowledge creation, and one in which epistemological issues loom large. So the first thing that I want to do here is to convince you that simulation is in fact a deeply creative source of scientific knowledge, and to give a taste of its complex and motley character.

The second thing I want to do is to argue that the complex and motley nature of this epistemology suggests that the end results of simulations often do not bear a simple, straightforward relation to the theoretical backgrounds from which they arise. Accordingly, I want to urge philosophers of science to examine more carefully the process by which general theories are applied. It is a relatively neglected aspect of scientific practice, but it plays a role that is often as crucial, as complex, and as creative as the areas of science philosophers have traditionally studied: theorizing and experimenting.

TWO

Sanctioning Models: Theories and Their Scope

Let us suppose that we are confronted with a physical system of which we would like to gain a better understanding: a severe storm, a gas jet, or the turbulent flow of water in a basin.[1] The system in question is made up of certain underlying components that fall under the domain of some basic theoretical principles. We begin by making two assumptions: we know what the physical components of the system are and how they are arranged, and we have great confidence in those principles.

The assumptions we have made so far about our system often allow us to write down a set of partial differential equations. In principal, such differential equations give us a great deal of information about the system. The problem is that when these underlying components of the system—whether they be solid particles, parcels of fluid, or other constitutive features—interact as we suppose they do in our physical system, the differential equations take on an unfortunate property. In the types of systems with which the simulation modeler is concerned, it is mathematically impossible to find an analytic solution to these equations—the model given by the equations is said to be analytically intractable. In other words, it is impossible to write down a set of closed-form equations. A closed-form solution to a set

---

1. See Wilhelmson 1989 on the simulation of a severe storm; Smarr 1985 and Kaufmann and Smarr 1993 on the simulation of intergalactic gas jets; and Moin and Kim 1997 on computer simulation in the study of turbulence.

of differential equations is a set of equations that are given in terms of known mathematical functions and whose partial derivatives are exactly the set of differential equations we are interested in.

These problems are not new to the computer age, nor are attempts to overcome them. In the past, modelers have focused their attempts on analytic techniques for finding approximate solutions to the differential equations in question. In many instances, they have successfully used these techniques to generate closed-form functions that are approximately valid. That is, for problem situations such as the three-body problem, modelers have found functions that can be shown to have the same qualitative character as the unknown solution to the equations to be solved.

But there are vast regions of possible solutions to interesting equations that are qualitatively different from any known closed-form function. To overcome this problem, the simulationist "discretizes" the equations and "solves" them by brute force. Discretization, in this case, is the process by which simulationists turn differential equations, which relate continuous rates of change over infinitesimal intervals, into difference equations, which relate rates of change over finite, or discrete, intervals. The values that these difference equations give can then be calculated by a digital computer, from one discrete moment in time to the next. This technique of simulation is often called "finite differencing."

Finite differencing is a discretization of both space and time. But discretization can occur in time but not in space: in this latter case, one uses a set of (complete in some norm) basis functions to transform the spatial part of the equations, so that the partial differential equation is turned into a set of (coupled) ordinary differential equations for the coefficients of the basis functions. This latter approach underlies methods, such as the finite element method, which are in common use in engineering applications.

Of course, the transformation of the differential equations into difference equations, or from partial differential equations to ordinary differential equations, constitutes an approximation. In principal, by choosing an appropriately "fine grid," that is, by using discrete intervals of space and time (or just of space) that are sufficiently small, simulationists can reduce the "damage" done by the approximation as much as they want. Unlike analytic techniques, which often require symmetry assumptions or the assumption of time independence, in principle, the simulationist using such methods need not necessarily impose such assumptions.

In practice, however, the amount of computer time and memory required to do these computations rapidly increases as the simulationist chooses smaller grids. To achieve results that are accurate enough to meet the needs of the simulation study, the problem posed frequently requires a grid that is too small for any reasonable allocation of computer time and memory. In such a case, the equations are said to be not only analytically unsolvable but also to be computationally intractable.

Simulationists deal with the problem of computational intractability by deviating from the most principled model suggested by theory. This can work in one of two ways. They can develop a model that is simplified and hence computationally cheaper than the theoretically principled one. This can allow for a finer grid. But in many cases a fine enough computational approximation is vastly out of reach or may not really even exist in principle. In such a case, they can supplement the model with features that have nothing to do with theory at all but are designed to compensate for the errors that the coarseness of the approximation is found to create. Either way, depending on what aspect of the solution the simulationist is interested in resolving, it is often advantageous to trade away theoretical rigor for expediency. The deciding factor is not which approach is most true to theory but which approach will produce the best solution set as the outcome of the simulation. The best solution set is the one that will best uncover or reveal the features of the system that are important for understanding it.

Keeping the existence of such strategies in mind, a number of interesting and interrelated issues arise. The first is the basic epistemological issue I raised earlier. Simulations are often used to reveal features of phenomena for which data are sparse. And so the theoretical ancestry of a simulation assumes a heavy duty in credentialing its results. How, then, do we evaluate the trustworthiness of simulations in such contexts when they tend to be so theoretically unprincipled? How do we tell, in other words, that the simulation is appropriately informative about the phenomena it is meant to investigate? This problem, which simulationists refer to as the problem of "validating" a simulation, is often kept conceptually distinct from the problem of verification. To verify a simulation, in the vernacular, is a mathematical issue. It means to establish that the results of the simulation really are informative about the mathematical solutions to the original model equations.[2]

2. There is a huge literature underlying verification and validation (V&V) in the engineering community, especially in areas that involve formal licensing requirements, such as in the nuclear engineering community; and there is a growing literature in areas such as the nuclear weapons stockpile program, where simulations are used to guarantee safety and functionality. For two extremely influential documents in this sphere, see Committee on the Evaluation of Quantification of Margins and Uncertainties Methodology for Assessing and Certifying the Reliability of the Nuclear Stockpile 2008 and American Institute of Aeronautics and Astronautics 1998. Many in the engineering

It is worth asking, however, if verification is really properly part of the activity of validation. Is it really the job of simulationists, in other words, to insure that their results conform to theory? So the epistemology of simulation is closely connected to a second question that I raised earlier. That question has to do with the relationship that the results of such simulations have to their theoretical ancestors. Should we think of them, as I asked above, as revealing the empirical content of those theories, or as something else entirely? How we answer the second question may have some bearing on our thinking about the first because it speaks directly to the question of whether or not verification and validation really are related in the way we often assume.

An Ideal Model and Its Limits

The first step in getting a handle on what an epistemology of simulation might be is to highlight and characterize the different inferential steps that take place during the process of simulation—those that might be subject to epistemic scrutiny. So I will make the point that a simulation study embodies a rich inferential process by outlining the essential steps involved in the study of complex phenomena using computational techniques.

If we look in a textbook on computer simulation, we are likely to find something like the conception of the steps involved that is illustrated in figure 2.1.

The idea is rather simple. We can illustrate it with a simple example: the pendulum. We begin with a *theory* that governs the phenomenon of interest. For a pendulum, that would be Newtonian mechanics. A basic understanding of the system suggests to us a *model* of the pendulum. Such a model would consist of an abstract description of the physical system (a point mass on the end of a massless string) and some differential equations, provided by theory, that describe the evolution of the values of the variables in the abstract model. The *treatment* consists of assigning values to basic parameters—in this case the length of the string, the mass of the bob, and the acceleration of gravity—and assigning initial values to the variables, which in this case would be an initial value for the angle of the pendulum and its angular velocity. Next, model and treatment are

community consider the latter of these to be the bible on V&V. Similar literature for the basic science community does not exist.

2.1 From theory to data in simulation.

combined to create a *solver*—a step-by-step computable algorithm that is designed to be an approximate, discrete substitute for the continuous differential equations of the model—and the solver is run on a computer to produce *results*.

Despite the simplicity of the example, this schema captures a great deal about many computer simulations of physical systems. But there are also several ways in which this ideal story of the construction of a simulation can and should be complicated. To gain some appreciation for this, let's look at a more complex and realistic example—a now famous project led by meteorologist Robert Wilhelmson: a computer simulation of a severe thunderstorm. The purpose of the simulation, wrote Wilhelmson, was to provide "improved understanding of severe storm structure and evolution" (Wilhelmson et al. 1990, 20). The simulation generated a four-hour period of "solution space" for a system of nine partial differential equations that describe the time evolution of the dependent variables of the model. The discrete data comprising this solution were then subjected to a variety of techniques of data visualization in order to resolve the water and ice structure inside of a storm, to be able to see how air moves and rotates in and around a storm, and to discern various physical processes that influence storm rotation near the ground.

Wilhelmson's simulation model was based on a system of nine partial differential equations. For initial conditions, the researchers used observed conditions from one vertical column of air in an actual pre-storm environment. The model was then initialized using horizontally homogeneous values for each of the nine variables of the simulations, with a temperature perturbation at the horizontal center of the storm to get things started.

In the language introduced above, what we have so far is a model, consisting primarily of the nine differential equations, and a treatment. The model, at least in part, is inspired by some basic physics of atmospheric dynamics. Next comes a "solver." I use scare quotes here because this is a somewhat misleading term. It implies that we should conceive of the simulationist as being able to do something directly akin to finding solutions to the differential equations of the model for the values specified in the treatment. In practice, the situation is far more complicated.

There are in fact two aspects to creating a computable algorithm out of the model and treatment. The first is the transformation of the continuous differential equations into discrete difference equations. But doing this in the most straightforward possible way can lead to an algorithm that is either too computationally expensive; worse, it can lead to an algorithm in which the approximations that constitute the transformation from continuous to discrete equations cause instabilities, serious round-off errors, or other liabilities that make the output unreliable.

As a consequence—and here we come to the second point—simulationists must often create algorithms that deviate substantially from the those that follow most straightforwardly from the theoretically principled model. There are two ways they can do this. If the problem is merely computational tractability, that can be overcome by simplifying the model: simulationists will ignore factors or influences from their computational models because limitations of computational power make their inclusion impractical. Similarly, they might remove degrees of freedom from the model, or make what are known to be unrealistic symmetry assumptions.

The second course of action is slightly more surprising. Often, the question of whether some particular aspect of a system under study is crucial to the system's dynamics is not even the issue. At times the simulationist is acutely aware of the important influence of one component of the dynamics, and yet it is simply impractical to include it in a full-blown simulation. In such situations the simulationist will resort to adding mathematical relationships to the solver that have no direct connection to the original differential equations of the model. These rough-and-ready, theoretically unprincipled model-building tools typically involve relatively simple mathematical relationships that are designed to approximately capture some physical effect in nature that may have been left out of the simulation for the sake of computational tractability. When coupled to the more theoretical equations of a simulation, they allow the simulation to produce results that are more realistic than they could have been without some consideration of that physical effect.

For example, here is Wilhelmson's description of his simulation: "A very simple model is used to account for the development of rain. In many studies such simple models are sufficient for studying storm dynamics. Although simple, they provide the key storm-driving forces, namely, warming due to the release of latent heat as water vapor condenses and cooling due to evaporation of cloud and rain drops in unsaturated regions" (Wilhelmson et al. 1990, 22). Even further from the ideal of figure 2.1, perhaps, is the fact that simulation models sometimes incorporate mathematical features that are cooked up to overcome what trial and error reveals to be problems with the discretization schemes of their original models. There are many examples. One rather well-known one occurs frequently in simulations involving the turbulent flow of fluids. It is sometimes referred to as *eddy viscosity*.

Again, the idea is rather simple. In turbulent fluids, vortices, also called eddies, form at an astonishing range of scales, from the very large, to the very small. These vortices play an important role in the transport and dissipation of energy and other physically meaningful variables. And they do so at all scales—from the scale of the system as whole all the way down to scales comparable to intermolecular distances. It is virtually impossible to create a discretization scheme that is fine enough to capture vortices, or eddies, all the way down to these smallest scales. So what a good simulation needs to do is to add a cooked-up piece of mathematics to the model that will do the same work that one imagines these eddies would do if they were being resolved at the finest scales. One such cooked up piece of mathematics is called eddy viscosity.

A nice example of the use of eddy viscosity is a simulation designed to study the convective properties of red giant stars (Porter, Anderson, and Woodward 1998; Jacobs, Porter, and Woodward 1998). A red giant star has a hot, dense stellar core that is surrounded by a very large and very rarified envelope. A typical red giant envelope is as large as the orbital radius of Jupiter. The stellar core of a red giant generates heat, which then moves to the surface primarily by means of convection. Since the diffusion of heat via radiation of light is not nearly as efficient at moving heat through the envelope as is convection, heat is moved through the system primarily by the process of gas being heated by contact with the hot core, which then becomes buoyant and rises upward.

Such stars are of special interest to nonlinear scientists because they are convectively unstable almost all the way through to the core. In younger stars, heat from nuclear fusion finds its way to the surface primarily through radiative diffusion. Only during the cooler last third of their journey is this heat transported by convective motions of stellar gas. In red giants, however, the complex and turbulent process of convection begins much nearer to the core, and so these stars exhibit particularly complex and unstable motions of fluid.

Modeling the convective movement of energy through this system is correspondingly difficult. Very small changes in temperature, pressure, and density in one part of the system may lead to turbulent vortices elsewhere, and small surface eddies can lead to large convective flows. Even the tiniest eddies, however, play a role in transporting these quantities around the system. Thus, if the model is to capture large-scale effects with any degree of accuracy, it must take into account the effects that take place on small scales—without requiring computations that overwhelm the computing system.

The basic equations that govern the model are the Euler equations for fluid dynamics. These are a relatively simple form of fluid dynamical equations. They are based on the laws of conservation of mass, momentum, and energy. They ignore viscosity but include the effects of compressibility (Porter and Woodward 1994). The use of the *inviscid* Euler equations is an ordinary idealization. Every fluid has some viscosity, but some fluid-flow problems can be treated inviscidly. In fact, viscosity contributes to the dynamics of the star in crucial ways. It does so, however, at length scales that are far too small to be tracked by a reasonable computer program. Researchers deal with this problem using eddy viscosity:

Viscous effects, which act only on tiny scales unresolvable by the computational grid, were approximated by a carefully formulated [eddy][3] viscosity. This viscosity of the numerical scheme dissipates kinetic energy of fluid motion into heat, like the real viscosity of the gas, but on the much larger scales of the computational grid. This [eddy] viscosity was carefully designed to restrict its dissipative effects to the shortest length scales possible, consistent with accurate representation of the nearly inviscid flow on the longer length scales. (PORTER, ANDERSON, AND WOODWARD 1998)

Eddy viscosity is a nice example of what is often called a *parameterization scheme*. Such schemes are elements of a simulation's algorithm that are designed to capture effects that slip between the cracks of the discretization grid. Parameterization schemes are extremely common in climatology, especially in global climate models.

Sub-grid processes are represented by *parameterisation schemes* describing their aggregated effect over a larger scale. These schemes are often referred to as "model physics" but are really based on physics-inspired statistical models describing the mean quantity in the grid box, given relevant input parameters. The parameterisation schemes are usually based on empirical data (e.g., field measurements making in-situ observations), and a typical example of a parameterisation scheme is the representation of clouds. (BENESTAD 2007)

3. In this quotation I have used the term *eddy viscosity* to replace their use of the term *numerical viscosity* because I think this is the now more standard use of the term. *Numerical viscosity* more standardly refers to a viscous-like effect that arises as an artifact of a poorly chosen numerical solver. Neither of these should be confused with *artificial viscosity*, which I discuss in later chapters.

The model of the star also needs to track how heat moves through the system via conduction. The modelers assume that the rate of thermal diffusivity depended only on the gas pressure, and could therefore be treated with relative simplicity. Finally, the model needs to account for how energy is lost from the surface of the star. The actual physics of this process are quite complicated, but the researchers were able to argue for a much simpler treatment of the problem. They simply used the standard formula for the radiation of a black body, and applied it exclusively to those parcels of fluid that, based on their calculated pressures, were likely to be found close enough to the surface to be able to efficiently radiate heat.

What this example illustrates quite nicely is the difference between the model you would write down from theory if you had no limitations to your computational power and the model you end up discretizing into an algorithm. And the difference does not just consist in the use of ordinary idealizations, at least as we ordinarily conceive of them. It also involves the substitution of phenomenological relationships for "real physics"— for example, substituting black-body radiation for the real physics of energy dissipation—and also the inclusion of cooked-up techniques like eddy viscosity to mitigate limitations in the differencing scheme.

Another nice example, again from climate science, of the kind of modeling methods I have in mind is the famous "Arakawa operator," which has been discussed by Johannes Lenhard (2007). In the 1950s, Norman Phillips built one of the first simulations of the circulation of the earth's atmosphere. The simulation was very successful for the first few weeks of simulation time, but after that it began to exhibit instabilities and eventually "blew up." In other words, it was an artifact of the numerical scheme that the amount of energy stored in the system began to grow exponentially after a certain period of time. This frustrated climate scientists for several years. A climate scientist named Akio Arakawa eventually arrived at the solution to the problem. Arakawa replaced the mathematically principled discretization of the basic equations of Phillips's model with his own technique. I quote Lenhard:

Arakawa did not follow the common practice of replacing the Jacobi operator with the discrete Jacobi operator, but used his own specially constructed Arakawa operator. What is decisive is that the Arakawa operator overcame the nonlinear instability and permitted a long-term stable integration. . . .

To achieve the stabilization of the simulation procedure, Arakawa had to introduce additional assumptions, *some of which ran directly counter to experience and to the laws underlying the theoretical model.* (188; MY EMPHASIS)

The point of these examples is to begin to complicate somewhat the ideal of simulation construction depicted in figure 2.1. According to that ideal, we have theory, or basic physics, guiding a choice of model, and then the selection of a numerical scheme to find solutions to the equations of the model. In these examples, however, we see two complications. The first is that theory is at best guiding, rather than determining, the choice of model. In fact, we can clearly see three distinct kinds of sources for the bits of mathematics that go into these models. The first is theory. The theory of fluid dynamics provides the Euler equations that go into the model of the star. The second we might call something like "physical intuition." It is not high theory but something more akin to physical intuition that guides Wilhelmson in his choice of mathematics for rain development. The same should be said of the choice of the standard black-body radiation formula for the dissipation of heat in Porter and Woodward's simulation of the star.

Finally, a third kind of consideration altogether guides the choice of a scheme like eddy viscosity or the Arakawa operator. We will discuss such techniques in greater detail in what follows, but for now let us simply remark that they come neither from theory nor from physical intuition, but from attempts to respond to the constraints imposed by the limitations of our computational abilities—often as those limitations are observed to emerge through trial and error. And this highlights a second fairly straightforward way in which the ideal of figure 2.1 is too simple. The arrows that point in the direction of determination in that figure all point to the right. But the choice of the Arakawa operator was guided by what was observed in the output of Phillips's original simulation. It is only after simulationists observe the artifacts generated by their original choices that they go back and tinker with their choice of the model to be discretized, attempting to get better results the next time. There is much more significant feedback from left to right than the ideal of figure 2.1 suggests.

There is at least one other respect in which figure 2.1 needs to be complicated that deserves mention. This has to do with the rightmost part of the figure. Once a simulation model is implemented on a computer in the form of a particular algorithm, the algorithm produces results in the form of a data set, often a very large one. But users of the simulation are rarely interested in this pile of numbers in such a form. And it is equally rare that they have *confidence* in this pile of numbers in all of its gory, undifferentiated detail. There is usually still much work to be done.

The data set requires interpretation. It can be visualized, subjected to mathematical analysis, and used in conjunction with other sources of knowledge, including observation, in order to arrive at the final goal of a simulation study—what I call a *model of the phenomena*. This, rather than a pile of numbers, is what a simulationist generally aims to produce.

A model of the phenomena is a picture of the phenomenon of interest that embodies all of the relevant knowledge about it, gathered from all the relevant sources (of which the simulation may be only one source). It can comprise mathematical relations and laws, images (both moving and still), and textual descriptions. Particularly in the sciences of complex nonlinear systems, the model of the phenomena is rarely a quantitative description of a single system under a single set of initial conditions. Rather, it represents an attempt to summarize the basic qualitative features of a whole class of structurally similar phenomena—and to single out the trustworthy ones from those that may be artifacts of the simulation. It might include such features as the following:

- An emergent, high-level mathematical relationship among certain aspects of the system, such as a scaling law.
- A transport mechanism: any effect, such as diffusion, turbulence, an instability, or viscosity, that explains the movement of some entity or quantity, such as mass, energy, or angular momentum, through a particular system.
- Threshold values of parameters; for example a Reynolds number at which a system undergoes the transition from soft to hard turbulence.
- Characteristic coherent structures (like the red spot of Jupiter).
- Characteristic geometries of flow.
- Patterns of interaction and competition among coherent structures.

In Wilhelmson's simulation of the severe storm, the data set generated by the simulation was composed of values for each of the nine dependent parameters at each of the points on the space time grid of the simulation. This data set was then subjected to a variety of complicated and labor-intensive visualization techniques designed to "reveal the inner dynamics" (Wilhelmson et al. 1990, 20) of the phenomena.

The ultimate goal was to produce a visual record of how the basic, internal, stable structural features of the storm evolve and to understand the internal mechanisms that are at work in creating and preserving the stability of these structures. The researchers generated images corresponding to naked-eye observations of the simulated storm as well as images corresponding to those generated by surface reflectivity radar. The visual viewpoint was generated by rendering images of the surfaces that enclose regions of cloud (small water droplets and ice particles) and regions of rain within the storm. This created a time series of images that depicted what the storm cloud would look like to the naked eye from some particular vantage point. Images traditionally generated from reflectivity radar are two-dimensional cross sections that are color-coded to graph the concentration of raindrops at every point in the cross section of the storm. The simulationists recreated these images from the data set generated by the model.

Next, they used imaging techniques to study the patterns and mechanisms of airflow inside the storm. Wilhelmson's team used the computed velocity field data in order to simulate the trajectories of imaginary "weightless tracer particles" through the storm environment. The team also used long streamers to display the trajectories of selected air particles inside the storm. These streamers allowed the researchers to view the major, stable, long-lived air currents.

Another important aspect of the flow is the vorticity. In particular, researchers are especially interested in depicting the patterns of streamwise vorticity, the rotation of air around an axis parallel to the direction of flow. For this purpose they used differently colored ribbons whose degree of twist is in proportion to the quantity of streamwise vorticity in that region of the flow line. All of these visualizations were preserved as both still images and full-motion video.

Once the researchers succeeded in visualizing these aspects of the flow, they were able to make use of these visual representations to identify some of the key structures and trajectories in the inner dynamics of the flow of air, ice, and water through the storm system. They were able to use this knowledge to construct "a model of storm evolution and persistence"(Wilhelmson et al. 1990)—a model of the phenomena for severe storms. Meteorologists researching storm dynamics are particularly interested in the question of how severe storms maintain their longevity and develop and maintain their rotational character. The researchers seek an answer to these questions by analyzing how the basic geometry of the main flow features works to create the features of the storm that are known to be important in preserving its basic structure.

A good example of this kind of explanation involves the updrafts in the storm and the vorticity of this flow. Wilhelmson and his colleagues were able to show that an updraft with a high degree of streamwise vorticity will become helical in character, and they have argued that this type of flow is essential for reducing the energy dissipation in a severe storm, thus prolonging its life. They identified four processes in the storm that contribute to vertical vorticity: advection (horizontal transport of air due to temperature variation), convergence of air, tilting of horizontal vortex lines into the vertical, and dissipation.

All of these remarks and examples are by way of complicating the ideal of figure 2.1, in which model construction is governed by theory. But in our examples, model construction is guided not only by theory, but also by physical intuitions, and by model-building tricks like eddy viscosity or the Arakawa operator that are tinkered into being through subtle feedback and interplay with the limitations of computational schemes that are revealed through trial and error.

In figure 2.1, the output of a simulation is a pile of unquestioned numbers. But in real examples of simulation, the numbers are only the first step in generating a model of the phenomenon. Wilhelmson's storm team could not begin to investigate the results of their simulations without first using the simulation's output to recreate images that are familiar to the meteorologist. The process of generating a model of the phenomenon involves sorting through the pile of numbers, trying to integrate it with other sources of knowledge, and determining which, if any, of the features revealed in that pile of numbers are trustworthy and reliable.

While figure 2.1 appears to be a fairly epistemologically straightforward process, the actual process that I have been describing is ripe with all sorts of uncertainties that need to be managed. The project of an epistemology of simulation is the study of the means by which we sanction belief in the outcome of simulation studies, despite this motley methodology. I want to argue, furthermore, that in order really to understand the relationship between models of phenomena and scientific theories—that is to say, broadly speaking, if we want to understand the relationship between scientific theories and their applications in contemporary science—we need to understand the processes by which these results get sanctioned.

Verification and Validation

If we turn once again to the kinds of remarks we are likely to find in the introduction of a textbook on computer simulation, or even to the technical literature of verification and validation, the following conception of the epistemology of simulation accompanies the ideal of figure 2.1. On this conception, the epistemology of simulation can be cleanly divided into two components: so-called verification and validation. *Verification*, on this conception, is the process of determining whether or not the output of the simulation approximates the true solutions to the differential equations of the original model. *Validation*, on the other hand, is the process of determining whether or not the chosen model is a good enough representation of the real-world system for the purpose of the simulation. Here is a characteristic passage: "Verification deals with mathematics and addresses the correctness of the numerical solution to a given model. Validation, on the other hand, deals with physics and addresses the appropriateness of the model in reproducing experimental data. Verification can be thought of as solving the chosen equations correctly, while validation is choosing the correct equations in the first place" (Roy 2005).

It is clear that this two-step conception of the epistemology of simulation fits nicely with the ideal of figure 2.1. In figure 2.1, first you choose a model on a principled basis, and then you try to solve the equations of that model. According to the conception embodied in V&V, the epistemology of simulation is cleanly divided along those lines. In validation, you have to determine whether you have chosen the right model. Since the model has been chosen in a principled way, you are supposed to be able to do this independently of the results you get out of your solver. In verification you have to determine whether you have found good solutions to that model. This is supposed to be principally a mathematical or computer science question. The question of the appropriateness of the model is conceptualized as being independent of the question of the fitness of the solver.

What I would like to argue here is that the epistemology of simulation does not divide as cleanly into verification and validation as this picture suggests. I would argue, that is, that simulationists are rarely in the position of being able to establish that their results bear some mathematical relationship to an antecedently chosen and theoretically defensible model. And they are also rarely in a position to give grounds that are independent of the results of their "solving" methods for the models they eventually end up using.

I think this claim is important for two reasons. The first reason is that when we look past the V&V model of the epistemology of simulation, we find an epistemology that is much closer to the epistemology of experiment than the V&V conception can really account for. I explore the consequences of this in detail in the following chapter. The second reason is that this issue of verification and validation dovetails nicely with the recent debates in the philosophy of science I discussed earlier about the scope of theories. The connection to this debate will become clear in what follows.

Let us first examine the claim that the epistemology of simulation divides cleanly into verification and validation. To do this, we have to look carefully at the epistemological strategies that are available to simulationists. What should be clear is that we should not expect to find anything like a logic of the epistemology of simulation modeling. What we can offer, instead, is something akin to what Allan Franklin provided in *The Neglect of Experiment* (Franklin 1986). Franklin asks, "How do we come to believe rationally in an experimental result?" His answer is that scientists use various strategies that philosophers ought to be able to see as providing grounds for rational belief in experimental results. While Franklin provides a list of twelve or so such strategies, he is careful to note that his list is neither exclusive nor exhaustive, and that no subset of the list is a necessary or sufficient condition for rational belief. He also notes that grounds for rational belief do not guarantee certain knowledge. Sometimes we may rationally believe something that we all later regard as wrong (Franklin 1986, 190–91).

The epistemology of simulation is an analogous project, but with an added benefit. Once we underscore the various strategies that simulationists use to provide grounds for belief in their results, we will not also gain a better understanding of what makes belief in the results of (some) simulations rational; we can only gain a better understanding of the relation between theories and their applications.

One of the central themes of Franklin's work is that experimenters are constantly preoccupied with scrutinizing experimental setups to uncover possible sources of artifact. Then they can work to eliminate the impact of these disturbances on experimental results. The same is certainly true of simulationists.

In what follows, I outline some of the strategies available to simulationists to argue for the reliability of their results. A central conclusion that I would like to draw from what follows is that these strategies are best understood as being aimed at providing grounds for belief that a simulation provides reliable information about the real-world system being simulated. They are not aimed at providing grounds that the computer is providing solutions to the original equations of the theoretically motivated model—at uncovering the empirical content of some theory.

To be sure, one of the principle kinds of arguments that the proponent of a simulation model offers in support of its adequacy is that the reliability of the computational techniques she has used are underwritten by sound mathematical theory and analysis. The simulationist, in other words, will certainly mount the best argument she can that the mathematical techniques used to turn the original model equations into the final algorithm are sound. But these arguments are invariably terribly weak. The history of computer simulations is littered with examples of computational methods that should have worked but did not because the numerical methods failed. So these kinds of arguments—the ones that are entirely mathematical in character—would never by themselves provide much rational ground for believing in the results of simulations. And there are good reasons for this—the arguments themselves often depend on linearity conditions that do not in fact apply to the conditions that are of interest to the simulationist.

Rather, a host of other strategies are required and in fact predominate. The most important strategy available to the simulationist, sometimes called "benchmarking" or "calibration," is to show that relevant output of the simulation matches what is known about the phenomena. Simulation models can be benchmarked in three different ways: by comparing their results to experiments, to analysis, and to other simulations. The first criterion that a simulation must meet is that it be able to reproduce known analytical results. Even for complex systems, there sometimes exist, under highly constrained conditions, limited analytical results for the full equations of a mechanical model. Typically, these results apply to highly symmetrical, equilibrium-state instances of the system, or to instances that can be studied as small deviations, or perturbations, of such instances.

Results can also be compared to the output of another simulation if it uses a different algorithm, or, even better, if it is of a small local region within the broader system and makes use of a more complete or refined model.

But the most important way in which simulation results are benchmarked is against real-world data. Simulationists expend a great deal of effort gathering data from experimental sources in order to benchmark their models. This kind of comparison is often not as easy as it might seem, since data from these different sources are hard to come by and may come in different forms. For example, simulation data and experimental data are not always obtained at the same spatial position within a system; the grid points of the simulation do not often correspond to the locations of probes within the experimental setup. Moreover, it is typically less interesting to compare detailed data—piece by piece—than it is to compare the characteristic features of the simulation results and empirical results, especially when the empirical results come in the form of flow visualization (Shirayama and Kuwahara 1990, 67).

Good benchmarking, therefore, requires the skilled judgment of a good observer; absent an observer who can compare images against images, there is no metric of similarity between the different data sets that need to be compared. Visualization is by far the most effective means of identifying characteristic features within complex dynamical data sets, and so it is the most effective means of judging the degree of benchmarking success a simulation enjoys against real data.

Besides benchmarking, there are other strategies available for sanctioning simulation results. One can check to make sure that, when simulated, the system responds as expected to changes in values of the parameters—a fluid-flow problem that becomes more laminar, for example, when viscosity is decreased, might not be being properly simulated, since we have systematic background knowledge that leads us to expect the opposite. One can also check to see if the simulation is capable of reproducing basic functional relationships that are predicted by higher level, or more phenomenological theories and laws.

I take it then, that we have found two important facts about simulations. The first is that simulationists often settle on models that they could never justify on a purely theoretically principled basis. There are, in the end, at best very weak theoretical arguments to be given for the ultimate choice of model. The second is that there are also, at best, weak mathematical arguments that can be given to support the claim that the output of a simulation approximates the mathematical content of those models. The sanctioning of simulations does not cleanly divide into verification and validation. In fact, simulation results are sanctioned all at once: simulationists try to maximize fidelity to theory, to mathematical rigor, to physical intuition, and to known empirical results. But it is the *simultaneous confluence* of these efforts, rather than the establishment of each one separately, that ultimately gives us confidence in the results. When simulationists argue for the trustworthiness of the models of phenomena they create under the guidance of theory, they offer no reasonably strong arguments that those models approximate the principled content of any particular theory.

It is true, of course, that simulationists do their best to show that their results are as close as possible to the real solutions of the equations that form the basis of their original models, and often they can offer very strong arguments to this affect for limited regions of solution space. The problem is that, in practice, the models they begin with are so complex, and rely so heavily on nonlinear equations, that the arguments they can offer for such conclusions are incredibly weak when they are applied to the entirety of the solution space. When models are sufficiently complex and nonlinear, it is rarely possible to offer mathematical arguments that show, with any degree of force, that verification is being achieved. What simulationists are forced to do is to focus, instead, on establishing that the *combined* effect of the models they begin with and the computational methods they employ provide results that are reliable enough for their intended purposes. This, of course, is hard. And it requires more finesse than we would expect if we thought that the activities of verification and validation could be kept entirely conceptually separate.

One can think of this, in a sense, as a kind of a Duhem problem. The Duhem problem is that when a prediction fails to match observed data, we do not know whether to blame the theory we were testing or an auxiliary hypothesis. Similarly, when a computational model fails to account for real data, we do not know whether to blame the underlying model or to blame the modeling assumptions used to transform the underlying model into a computationally tractable algorithm. But to the extent that the Duhem problem is a problem about falsification—about where to assign blame when things go wrong—the present problem is more than a Duhem problem. When a computational model succeeds—when it provides results that are adequate for a particular purpose—it might not in fact be because either the underlying model is ideal or because the algorithm in question finds solutions to that underlying model. It might rather be because of what simulationists sometimes call a "balance of approximations." This is likely the case when a model is deliberately tailored to counterbalance what are known to be limitations in the schemas used to transform the model into an algorithm. When success is achieved by virtue of this kind of back-and-forth, trial-and-error, piecemeal adjustment, it is hard even to know what it means to say that a model is separately verified and validated.

I should be clear about one thing here: I am not arguing that verification and validation are not separable activities in practice. I am not urging practitioners of simulation-model evaluation to abandon this useful conceptual distinction. Rather, I would prefer to be read as urging philosophers not to overinterpret the fact that this pragmatic distinction exists among practitioners. I do not want philosophers to conclude from the fact that practitioners distinguish validations and verification that the principal work of evaluating a simulation is a mathematical, as opposed to an empirical matter—at least not in all cases. What is a pragmatically useful distinction for practitioners can be misleading to philosophers if it is taken to be more conceptually clear than it is.

And the apparent conceptual clarity afforded by keeping these two activities distinct is misleading. It can make it seem, for example, that it is precisely the activity of verification that is unique to simulation, as opposed to other modeling activities where the conclusions one can draw from the model can be inferred analytically. And finally, it might seem that questions of verification are entirely mathematical questions, without philosophical meat. When, for example, Roman Frigg and Julien Reiss (2009) argue that simulation is epistemologically uninteresting precisely because its only epistemologically unique feature is entirely mathematical in nature, they are making just this mistake.

Uncovering the Content of Theories

If verification and validation are often carried in tandem, rather than being clearly conceptually distinguishable, then this raises a deep question about the relation of models of phenomena to theories: Should we indeed think of the models of phenomena that simulationists provide as approximate qualitative characterizations of the empirical content of some theory—as the ideal, undisplayable "solutions" of the theoretical principled equations? Or is it possible, on the other hand, that theories can be used to guide the creation of representations of phenomena that are not, properly speaking, part of the "content"[4] of those theories? This question is of course closely intertwined with the epistemological conclusion above.

Closely intertwined, but not identical. If we are interested, after all, in the relation that models of phenomena have to theories, we need to distinguish two different issues. The first is whether simulationists offer direct arguments that their models of phenomena approximate the principled content of some theory. And the second is whether those models actually approximate that content. In fact, one could very well argue that when simulationists argue for the trustworthiness of their models of phenomena—when they argue that those models approximately depict the behavior of some real-world system—they are providing an indirect argument that the models approximately uncover the content of the best theory of that real-world system. After all, if the theories that guide the construction of the simulation model are empirically adequate, in the way that philosophers traditionally understand that term,[5] then an argument for agreement with real-world behavior would be a de facto

---

4. In the sense, for example, that participants in the debates on the structure of theories intended (see, e.g., Suppes 1962; Suppe 1974, 221–30; Van Fraassen 1970 and 1980, 64–70).

5. For Van Fraassen (1970), for example, a theory is empirically adequate if one of the family of models that the theory comprises is isomorphic to all possible observations that fall under the purported domain of the theory.

argument for agreement with the empirical content of any relevant empirically adequate theory.

But the above argument makes a fairly strong assumption about theories and their empirical adequacy. It assumes that the theories that guide the construction of the relevant models have wide scope. It assumes that, despite our own cognitive limitations, *in principle*, the scope of every theory by itself is always as wide as the scope of the set of models we are capable of building under the guidance of that theory.

Specifically, the above argument assumes that simulations can never produce good results that fall outside of the principled scope of the theories that originally guide their construction. This assumption should not, I think, be taken for granted. In fact, it is closely connected to the debate I mentioned earlier about the scope of theories. The debate, we should recall, is about whether laws and theories in science are tightly restricted with respect to the features of the world that fall under their domain—or whether, on the other hand, it is the nature of fundamental theories that they have universal domains.

The question that concerns us here is whether it is possible that simulations can sometimes produce results that fall outside of the principled scope of the theories that originally guide their construction. This does not speak precisely to the debate between the fundamentalist and the pluralist, since the simulations we are looking at are not guided by *fundamental* theory. But it speaks to a very closely related question: whether theoretically guided model-building methods sometimes generate genuinely novel knowledge—novel in the sense of not being even "implicitly contained" in preexisting theory, whether or not it is generally the case that the principled scope of every theory extends, even implicitly, as far as all of its nonprincipled applications. In fact, since it is not about fundamental theory, an affirmative answer to this question would be a weaker claim than the one defended by Cartwright.

Still, I do not offer a compelling argument for even the weaker claim here. We are talking, after all, about the nature of areas of solution space that are, in principle, inaccessible. Compelling arguments for either one or the other are probably not forthcoming. Indeed, it is hard to say what the criteria ought to be for resolving debates about the scope of theories when principled models and their solutions do not exist.

What I do want to urge, however, is that we resist the temptation to take the opposite claim for granted: that a simulation can only be useful and sanctionable if the systems it represents fall under the principled scope of its theoretical ancestors. And there are two reasons to resist that temptation.

The first reason is that a dogmatic commitment to this expansive conception of the scope of theories obscures scientific practice. It inhibits us from seeing how theory guides *but does not determine* how the models of phenomena for complex systems are constructed. It obscures the rich epistemology of simulation and reinforces the misleading conception that epistemology can be neatly divided into settled theoretical issues on the one hand (validation) and mathematical/computational ones on the other (verification). The epistemology of simulation, as we have seen at length, is very much an empirical epistemology and not merely a mathematical or logical one. When we turn, in the next chapter, to examining the epistemological connections between simulation and experiment, it will be an impediment if we insist on seeing simulation as the process of uncovering the empirical content of theories.

The second reason is that there are principled reasons to doubt the claim in the first place. Consider the case of simulations of fluid-flow problems, which are generally based on a form of the Navier-Stokes equations. Analysis of the three-dimensional form of these equations has never established the formal uniqueness and existence of the right kinds of solutions. There is even some evidence that singularities might almost inevitably form, which would imply a breakdown of the equations. Nevertheless, it seems quite clear that, at least for certain kinds of problems, these equations can be used to guide the construction of very successful simulations.

In later chapters, we take a close look at so-called multiscale simulations. These are simulations that divide and conquer the problem they are designed to solve by using different theoretical frameworks to guide model construction in different regions of interest. The models of each of these separate regions are sewn together into one unified simulation with "handshaking algorithms." It is doubtful, therefore—or at least there is no particularly good reason to believe—that the net output of these simulations is an approximate solution to any particular set of theoretically motivated equations.

In the end, the debate about the scope of theories is one for which, until we reach the end of enquiry, the criteria for its resolution are not obvious. But whatever metaphysical position we hold in that debate, there are important and challenging epistemological and methodological issues in scientific theorizing that will be overlooked if we see theories as fully articulated structures and treat calculational problems as *merely* the result of practical limitations.

When it comes to complex systems, we simply cannot bend our theories to our cognitive will—they will not yield results with any mechanical turn of a crank. The models that we need to construct in order to do our science need to be constructed delicately and from as many sources as are available. Consequently, these models are best viewed not as mere solutions to theoretical equations; they are rich, physical constructs that mediate between our theories and the world.

THREE

Methodology for a Virtual World

In the last chapter, I argued that we should resist the temptation to view the results of simulations, even theoretically motivated simulations, as part of the proper empirical content of theories. I suggested, furthermore, that one reason we should resist that temptation is because it obscures what I called the "empirical" character of the epistemology of simulation, implying that the latter has close connections to the epistemology of experiment. It is time to make some of these claims a bit more precise.

One need not look very far into the literature of simulation to find that it is full of language that strongly evokes metaphors of experimentation. The interpretation of metaphorical language, however, requires some care. It can be nothing more than loose metaphor. We have to remember, moreover, the important role that well-established theory plays in most simulations, especially in the physical sciences. In such contexts, simulation is only possible precisely because good theories, well confirmed by a history of experiments, exist to underwrite them. If the metaphors are to be taken seriously, simulations would then seem to have the character of both experimentation and theorizing—two activities that are traditionally taken to occupy rather different spheres of scientific activity. Where, then, "on the methodological map,"[1] do techniques of computer simulation lie?

1. Galison 1996, 120.

One version of that question that we might ask is whether or not simulations are, properly speaking, themselves a class of experiments—or if not, what, if any, essential characteristics distinguish simulations and experiments. I address such questions in the next chapter. Here I want to answer the following questions: What epistemological features do simulations actually share with ordinary laboratory experiments? What can we learn from the epistemology of experiment—a more developed area of the philosophy of science—that we can apply to the epistemology of simulation?

If simulations that are underwritten by well-confirmed theory share important and interesting epistemological features with experiments, then we may be able to draw on some of the recent interesting work in the philosophy of experiment to illuminate our understanding of the application of theories to local models. So the question is really this: What can we learn about the epistemology of the construction of local models by drawing comparisons between the epistemology of experiment and that of simulation? That, in turn, could further amplify some of the lessons of the last chapter, as well as amplify and illuminate Mary Morgan and Margaret Morrison's recent discussion of the application of theories—in particular their characterization of "autonomous models."

> One of the points we want to stress is that when one looks at examples of different ways that models function, we see that they occupy an autonomous role in scientific work. We want to outline . . . an account of models as *autonomous agents*, and to show how they function as *instruments* of investigation. . . . It is precisely because the models are partially independent of both theories and the world that they have this autonomous component and so can be used as instruments of exploration in both domains. (MORGAN AND MORRISON 1999, 10; EMPHASIS IN ORIGINAL)

For our purposes, the term *autonomous models* is somewhat misleading. A better term would be *semiautonomous*. The claim frequently made by Morrison and Morgan that models are autonomous or independent of theory is meant to emphasize the fact that there is no algorithm for reading models off from theory. As I have emphasized, this is especially true of the models that drive simulations. While these models generally incorporate a great deal of the theory or theories with which they are connected, they are usually fashioned by appeal to, by inspiration from, and with the use of material from, an astonishingly large range of sources: empirical data, mechanical models, calculational techniques (from the exact to the outrageously inexact), metaphor, and intuition. In the end, the model that is used to run the simulation is an offspring of the theory, but it is a mongrel offspring. It is also substantially shaped by the exigencies of practical computational limitations and by information from a wide range of other sources. Still, to call these models completely autonomous, at least in this context, is to deny their obvious and strong connections to theory. This feature of simulation models—their semiautonomy—will be crucial to keep in mind if we want to understand how simulation can share aspects of theorizing and experimenting. And that, in turn, will be crucial if we want to better understand how theory guides the construction of local models.

It's All Metaphorical

We can start by looking at what some recent commenters on simulation have said about the relation between simulation and experiment. One view suggests that talk of "simulation" and "numerical experiments" is purely hyperbolic or metaphorical—simulation is nothing more and nothing less than using brute-force computational means to solve analytically intractable equations. A second view, in which the terms *simulation* and *numerical experiment* are taken quite literally, a simulation is a stand-in, or mimic, of a real-world system, and we can therefore perform experiments on it just as we can on any other experimental target. Finally, there is the view that simulation is a brand new "third mode" of science, neither experimental nor theoretical. In what follows, I weigh the merits of these three views, emphasizing how much each of them can contribute to understanding how models are to be evaluated.

Let's begin with the view that simulation is just a fancy word for using brute-force computational means to solve analytically intractable equations—and that all the experimental metaphors built into the language of simulation are nothing but empty metaphor. One perspective from which to see clearly that this view is wide of the mark has been offered by the philosopher R. I. G. Hughes, who has argued for a distinction between "the use of computer techniques to perform calculations" on the one hand and genuine computer simulation on the other (Hughes 1999, 128). For Hughes, the distinction hinges on the "thoroughly realist mode of description" that is used to describe the results of genuine simulations, and on the images that simulation produces, which often resemble photographs of material physical models. Thus, what distinguishes genuine simulations from mere number crunching is that simulations have genuinely "mimetic" characteristics.

There is certainly something right about this. The mimetic characteristics of an algorithm—one that uses sophisticated graphics and that is treated realistically by its users—are surely an important consideration that contributes, at the very least, to experimental metaphors concerning simulation. So psychologically, at the very least, working with a simulation is much more like doing an experiment if the simulation produces lifelike images reminiscent of laboratory photographs.

Of course, we are looking for something deeper than a superficial, psychological resemblance. In fact, if this were all that simulations, or "numerical experiments," had in common with experiments, then one might indeed be tempted to agree that talk of simulation as numerical experimentation was no more than mere hyperbole. To illustrate this point, we need look no further than the example of mere number crunching offered by Hughes; the use of a computer to calculate the orbits of planets in a three-body problem.[2] In such a case, nothing at all prevents creators of such an algorithm from imbuing it with high-quality graphical output, and from characterizing their results in as realistic a mode of description as they like. Since no distinction of great philosophical import could hinge on whether or not "mere" computation is embellished with graphic presentation, we must do more than cite the fact that some algorithms use graphics and some do not.

There are in fact two characteristics of true simulations, characteristics we are familiar with from the last chapter, that meaningfully distinguish them from mere brute-force computation in ways that connect them to experimental practice in an interesting fashion.

- Successful simulation studies do more than compute numbers. They make use of a variety of techniques (most of which, pace Hughes, involve imaging in one way or another) to draw inferences from these numbers.
- Simulations make creative use of calculational techniques that can only be motivated extra-mathematically and extra-theoretically. As such, unlike the results of simple computations that can be carried out on a computer, the results of simulations are not automatically reliable. Much effort and expertise goes into deciding which simulation results are reliable and which are not.

Let me elaborate on these two themes. First, simulations are interestingly like experiments because they involve data analysis. Suppose that we are confronted with a difficult fluid-flow problem. In such a case,

2. It is worth noting that the problem of calculating the orbits of multiple massive bodies mutually interacting via the force of gravity is highly nontrivial, and it is a mistake, in any case, to view it as mere number crunching. There are basic questions one can ask about solar systems (e.g., about their stability) for which it is not at all clear that any particular simulation method guarantees correct results.

we think that we know the governing principles (in this case, a form of the Euler equations), but we still do not know what kind of local behavior these principles predict. One approach would be to use a tractable, approximative, analytical technique. For example, analysts might use a series expansion, truncate all but two or three of the terms, and end up with an equation for which one can write a solution. Or one might use a finite-difference simulation technique. Both of these techniques employ approximations, both involve creative work, and both will, in turn, give rise to issues of justification.

One important difference between the two methods, however, is this: the analytic method will produce, as its result, an algebraic expression. That expression, in turn, can represent the behavior of a general class of systems. Various functional dependencies and patterns of behavior can easily be read off from a closed-form expression.

Numerical methods, on the other hand, provide as their results a big pile of numbers. If simulationists want to learn about the general qualitative features of a class of systems, then they must apply all the usual tools of experimental science for analyzing data: visualization, statistics, data mining, and so on. If they want to discover functional dependencies, they must also run a barrage of trials, looking at the results across a wide range of parameters. This aspect of simulation certainly carries the most obvious methodological characteristics of experimental work. It is the need, furthermore, to draw inferences from these piles of data that brings about what Hughes calls the "mimetic" aspects of simulation—their "thoroughly realistic mode of description."

In a review article on computational methods, a group of researchers describe the situation this way:

> The most common method for observing the behavior of laboratory flows is to make photographs using a variety of techniques that bring out specific features. . . . Computational fluid dynamicists naturally want to use similar techniques to display their results. Displays not only make comparisons with laboratory data much easier, but also are useful at getting at the fundamental physics of wave interactions, surface instabilities, vortex generation and other phenomena that may be involved in the flow.
> (WINKLER ET AL. 1987, 29)

Recall Hughes' insistence that genuine simulations be distinguished from mere step-by-step calculations on the basis of the former's mimetic qualities. It seems to me that what underlies this distinction is really the difference between computational methods that are used to make specific quantitative predictions (for example, where will the fourth planet in a five-body system be after 72 days?) on the one hand and computational methods used to "get at the fundamental physics"[3] on the other. It is only in the latter case that imaging techniques become necessary and that the computational model takes on a mimetic quality, so that techniques familiar to the laboratory observer can be applied. But whether or not imaging techniques per se are used, simulations that aim to get at global, diachronic features of the systems they model will invariably not only produce numbers but also draw inferences from those numbers in a manner that is richly analogous to experimental data analysis.

The second feature of simulations with respect to which they have a great deal in common with experimental practice is the constant concern with uncertainty and error. As I have already emphasized, although simulation often is initially motivated by well-established theory, the end model that drives the computations generally incorporates modeling assumptions that are not theoretically motivated. The results of the simulation, therefore, do not automatically come with a stamp of approval that carries the full faith and credit of the governing theory's epistemic credentials. According to Winkler, "Unless uncertainties are kept under control, the computational approach cannot uncover *new* physical phenomena" (1987, 29; emphasis in original).

That simulations and laboratory experiments both require the management of uncertainties is perhaps interesting in and of itself. But if we are interested in epistemological issues surrounding autonomous models, the really interesting question is whether or not they do so in analogous ways. If they do, then perhaps some of the insights from recent work in the philosophy of experiment will shed some light on the sanctioning of models.

I return to these questions later on. For now it is enough to address the challenge of this section with the following point. While having a mimetic quality is not in itself what gives a simulation interesting methodological and epistemological features, it is an important sign of other features that do. The extensive use of realistic images in simulation is a stepping stone that simulationists use in order to make inferences from their data. It is also a tool they use in order to draw comparisons between simulation results and real systems, which is part of the process of sanc-

---

3. Obviously, Winkler et al. are using the expression *fundamental physics* in a manner that is quite different than the one to which philosophers might be accustomed. I take it that they are using the expression to signify not the fundamentals laws or theories of the system but the emergent structural features of the dynamics like vortices, waves, and surfaces and their interactions—the important, dynamically significant, long-term and long-range features of the dynamics that give insight into its time-evolution.

tioning their results. It is the drawing of inferences and sanctioning of results that give rise to interesting philosophical connections between simulation and experimental practice.

The Computer as Experimental Target

Another view on how to locate simulation methodologically has been to interpret expressions like *simulation* and *numerical experiment* literally. The idea here is to interpret the simulation algorithm as literally "mimicking" the system or systems of interest and to understand what scientists do as performing experiments on the computer or computer algorithm, which acts as a stand-in for, or probe of, the system in question. Computer simulations are experiments, and the computer is the target of the experiment.

Practitioners of simulation often emphasize this point of view themselves, especially in popular or semipopular presentations that are geared toward expounding the virtues of numerical methods. For example,

A simulation that accurately mimics a complex phenomenon contains a wealth of information about that phenomenon. Variables such as temperature, pressure, humidity, and wind velocity are evaluated at thousands of points by the supercomputer as it simulates the development of a storm, for example. Such data, which far exceed anything that could be gained from launching a fleet of weather balloons, reveals intimate details of what is going on in the storm cloud. (KAUFMANN AND SMARR 1993, 4)

The germ of this idea probably comes from the father of computational methods himself, John von Neumann—although he actually expressed the idea in reverse. Working on highly intractable fluid dynamical problems at Los Alamos, von Neumann lamented the fact that he and his colleagues often had to perform difficult experiments just to determine facts that should, in principle, be derivable from well-known underlying principles and governing equations: "The purpose of the experiment is not to verify a proposed theory but to replace a computation from an unquestioned theory by direct measurement. . . . Thus wind tunnels are used . . . as computing devices . . . to integrate the nonlinear partial differential equations of fluid dynamics" (quoted in Winkler et al. 1987, 28). Once von Neumann's dream became ostensibly realized, and the wind tunnel was replaced by the computer, it is not altogether unnatural to view the resulting activity as performing experiments in a virtual wind tunnel.

Something like this view of computer simulation has received most attention from philosophers of science. This is principally a metaphysical claim about the relation between simulation and experiment—it is the claim that simulations are experiments—and we give this issue more attention in the next chapter. Here we are interested in epistemological connections. But it will help us to take a preliminary look at the view, since it sometimes takes on, or at least flirts with, an epistemological formulation.

Several philosophers have argued that simulations are experiments. Paul Humphreys (1994), for example, argued for a view like this vis-à-vis Monte Carlo simulations, on the grounds that when a program runs on a digital computer, the resulting apparatus is a physical system. Any runs of the algorithm, therefore, are just experimental trials on a physical target. This is also roughly the view that is espoused by Hughes. According to Hughes, there is no reason to resist thinking of computer simulations as experiments, since they lie on a "slippery slope" that makes them conceptually inseparable from experiment. The slope looks something like this:

1. We experiment on a model water wheel in order to learn about water wheels in general.
2. We experiment on an electrical damped harmonic oscillator in order to learn about mechanical damped harmonic oscillators that are governed by a structurally identical equation.
3. We run experiments on cellular automata machines in order to learn about systems with identical "symmetries and topologies"
4. Why not also say, then, that we perform experiments on computers running algorithms designed to simulate complex physical systems? (HUGHES 1999, 138)

It is not clear what exactly the epistemological content of these claims is meant to be. But Steve Norton and Frederick Suppe have explicitly argued that we can best understand the epistemological foundations of simulation by seeing it as a form of experiment—where the computer is the physical object being experimented on. According to Norton and Suppe, a valid simulation is one in which certain formal relations hold between a base model, the modeled physical system itself, and the computer running the algorithm. (See Norton and Suppe 2001 for details.) The relation in question is that of *realization*, where a system S_1 *realizes* a system S_2 just in case there is a many-one, behavior-preserving mapping from the states of S_2 *onto* the states of S_1. When the proper conditions are met, a simulation "can be used as an instrument for probing or detecting real world phenomena. Empirical data about real phenomena are produced under conditions of experimental control" (Norton and Suppe 2001, 89). *"Simulation modeling is just another form of experimentation, and simulation results are nothing other than models of data"* (92; emphasis in original).

In general then, the claim being made by Hughes and by Norton and Suppe is that the methodological structure of simulation is like that of experimentation because simulation proceeds in the following way:

1. Create an algorithm that accurately mimics a physical system of interest.
2. Implement the algorithm on a digital computer.
3. Perform experiments on the computer, which will tell us about the system in question—in the same way that experiments on an electrical damped harmonic oscillator tell us about mechanical damped harmonic oscillators.

Here we need to carefully distinguish two separate issues. The first is the extent to which, following Hughes, simulation can be seen as continuous with experiment. We address this issue in the next chapter. The second is the extent to which we can use what we know about the epistemology of experiment to understand the epistemology of simulation. The real problem with this sort of story, from the point of view of the latter issue—from the point of view of the project of this chapter—is that it begs the question of whether or not, to what extent, and under what conditions a simulation *reliably* mimics the physical system of interest. We are interested in applying some of the insights of the philosophy of experiment to methodological and epistemological issues vis-à-vis simulations. But to identify the methods of simulation with the methods of experiment in this way is to tuck away all of these important questions into step 1 and then to focus exclusively on step 3 as the step where the connection lies between simulation and experiment.

Hughes himself at least partially recognizes this problem. His solution is to say that computer "experiments" reveal information about actual, possible, or impossible worlds. To know that we are finding out about actual worlds, according to Hughes, requires an extra step. "Lacking other data, we can never evaluate the information that these experiments provide" (Hughes 1999, 142). This is not so much a problem for Hughes since his analysis is not intended to be epistemological. In fact, in the last quoted passage, Hughes is essentially disavowing any epistemological force to his analogy. But we must face this issue if we want to offer the analogy between simulation and experiment as a way of understanding what makes simulation results reliable. One of the central epistemological questions about simulation is "How do we evaluate the information that these simulations provide when other data *are* lacking?"

This state of affairs is reminiscent of a passage of Wittgenstein's, in which he is criticizing Ramsey's theory of identity:

> Ramsey's theory of identity makes the mistake that would be made by someone who said that you could use a painting as a mirror as well, even if only for a single posture. If we say this, we overlook that what is essential to a mirror is precisely that you can infer from it the posture of a body in front of it, whereas in the case of the painting you have to know that the postures tally before you can construe the picture as a mirror image.
> (WITTGENSTEIN, QUOTED IN MARION 1998, 70)

To adopt, without further comment, the analogy between simulation and experiment would be to make the same mistake as the one identified by Wittgenstein. Simulation is a technique that begins with well-established theoretical principles and, through a carefully crafted process, creates new descriptions of the systems governed by those principles. It is a technique that, when properly used, provides information about systems for which previous experimental data is scarce. Paraphrasing Wittgenstein, "What is essential to a [simulation] is precisely that you *can* infer from it" what some system in the real world is like, even when other data *are* lacking. If our analysis of simulation takes it to be a method that essentially begins with an algorithm antecedently taken to accurately mimic the system in question, then we have begged the question as to whether and how simulations can, and often do, provide us with genuinely new, previously unknown knowledge about the systems being simulated. It would be as mysterious as if we could use portraits in order to learn new facts about the postures of our bodies in the way that Wittgenstein describes. We need to understand how, as Kaufmann and Smarr (1993) suggest, we can reliably learn about storms from simulations, even when data about such storms are conspicuously sparse.

There is another deep problem with the emphasis on the mimetic qualities of simulation. Simulations often yield sanctioned and reliable new knowledge of systems even when nothing like the stringent conditions required by Norton and Suppe are in place. In practice, simulationists need not suppose—nor even begin to suppose—that their simulations perfectly mimic any particular physical system in order to convince themselves that certain qualitative properties of their results can reliably be attributed to the system being studied. So, while it is true that simulations are used to stand in for the systems they simulate, the relation between the simulation and the system is far more complicated than mimicry.

If there is a useful analogy to be made by philosophers between simulation and experiment, then that analogy ought to help make methodological and epistemological connections. These connections, in turn, should help us to apply some of the insights of recent work in the philosophy of experiment in order to gain an understanding of the conditions under which we should take simulation results to be accurate representations of real systems. Thus far, this proposal fails to do this because it assigns experimental qualities only to those aspects of simulation reasoning that occur *after* it is assumed that the simulation algorithm "realizes" the system of interest.

A Third Mode

A third view on simulation's methodological geography is that simulation represents an entirely new mode of scientific activity—one that lies between theory and experiment. For example, according to Fritz Rohrlich, a physicist and philosopher of science, "Computer simulation provides . . . a qualitatively new and different methodology for the physical sciences. . . . This methodology lies somewhere intermediate between traditional theoretical physical science, and its empirical methods of experimentation and observation" (1991, 507).

Historians, sociologists, and leading practitioners of the techniques themselves have expressed this view. In an introductory essay for a special issue of *Physics Today* on the subject of computational physics, Norman Zabusky wrote, "Supercomputers with ultrafast, interactive visualization peripherals have come of age and provide a mode of working that is coequal with laboratory experiments and observations, and with theory and analysis" (1987, 25).

Deb Dowling, a sociologist of science, has suggested a similar interpretation of simulation. She observes that simulation is like theory in that it involves "manipulating equations" and "developing ideas" but is like experiment in that it involves "fiddling with machines," "trying things out," and "watching to see what happens" (Dowling 1999, 264). The historian Peter Galison takes a similar view: "[Simulation] ushered physics into a place paradoxically dislocated from the traditional reality that borrowed from both the experimental and theoretical domains, bound these borrowings together, and used the resulting bricolage to create a marginalized nether land that was at once nowhere and everywhere on the usual methodological map" (Galison 1996, 120).

Locating simulation as lying *between* theory and experiment provides a natural perspective for both historians and sociologists. In Galison's case, it is a useful way of expressing how simulation could provide a trading zone between experimentalists and theoreticians and across disciplines. Dowling, in turn, has argued that simulation's ambiguity with respect to the experiment/theory dichotomy can play important social and rhetorical functions.

From the point of view of the project of this chapter, however, it is not clear what we gain by saying that simulation "lies between theory and experiment." What is of interest philosophically is to understand exactly what epistemological features of experimentation are shared by simulation and to use that to gain an understanding of the construction of local models. For these purposes, making simulation out to be an entirely new methodology and urging that it lies between theory and experiment is, at best, a good place to start.

Speculation versus Calculation

One way in which we might be able to make this approach do more work for us is to be clearer about the concept of "theorizing" when we say, for example, that simulation is an activity that lies between theorizing and experimenting. A good place to start is with Ian Hacking's repudiation of the traditional dichotomy of theory and experiment. He urges that we replace it with a tripartite division: speculation, calculation, and experimentation: "[By calculation] I do not mean mere computation, but the mathematical alteration of a given speculation, so that it brings it into closer resonance with the world" (Hacking 1983, 213–14).

The point that Hacking is making with his neologisms is that there are really two quite distinct activities that we often naively lump together under the label of "theorizing." The first activity is that of laying out basic theoretical principles: Maxwell's equations, Newton's laws of motion, Einstein's field equations, and so on. This is the activity Hacking calls "speculation." The second activity is what Thomas Kuhn long ago called "theory articulation." This is the hard work of making the aforementioned theoretical principles apply to the local, concrete systems that make up the real world. Hacking calls this activity "calculation," but I prefer the simple expression "model building" to Hacking and Kuhn's phrases because it emphasizes that this is an activity that often brings us *beyond* the original theoretical principles themselves. The expression "model building" also is meant to emphasize that the model being built is not properly seen as a component of the theory proper. The models constructed are offspring of theories, but they emerge as semiautonomous agents. In this respect, Hacking's pair of terms, *speculation* and *calculation,* are misleading, since in conjunction they suggest that there is nothing speculative about the process of building models under the guidance of theory —and nothing could be further from the truth.

The word *theorizing,* as it is *naively* used, expresses some amalgam of these two distinct activities, and it effectively collapses a valuable distinction. This sloppiness of language has persisted, in no small part, because most commentators on science, especially philosophers, have woefully underestimated the importance of theory articulation, or model building.

It is this inadequate appreciation of the importance of model building that Nancy Cartwright has labeled the "vending machine view of theories" (Cartwright 1999, 179–83). On this view, criticized by Cartwright, the ability of a theory to represent the world is captured precisely by the set of those conclusions that can be drawn deductively from the theory—drawn, moreover, with the ease with which we can extract candy from a vending machine. For Cartwright, theories do not by themselves have the power necessary to represent real, local states of affairs. Only what she calls "representative models" are fully able to "represent what happens, and in what circumstances" (180). It is these models that represent "the real arrangements of affairs that take place in the world" (180). On her view, the process of creating representative models from theory is complex and creative.

Widening the field to three activities instead of the traditional two makes it easy to argue that simulation is a form of theory articulation, or "model building." The models of phenomena that are the end products of simulation are strongly influenced by theory, but they are not part of the proper content of any theory. The fact that simulation work is such a creative and such an epistemologically delicate process is grist for Cartwright's mill. The difficulty of obtaining reliable simulation results testifies to the claim that theories do not dispense representative models as easily and conveniently as a vending machine dispenses candy.

Nevertheless, I do not want simply to make the point that simulation—rather than being either theorizing, or experimenting, or occupying some midway point between the two—is actually a form of what Hacking calls "calculation." Even with a tripartite distinction in hand, there are still aspects of simulation that, both methodologically and epistemologically, seem to have characteristics more commonly associated with experimental practice than with the pencil-and-paper varieties of model building. Nevertheless, I think it is very important to keep a keen eye on the distinction between the laying out of theoretical principles and the construction of local models when we throw around the word *theorizing*.

In fact, I would argue, something like what Hacking and Cartwright say about theories and models must be true if anything at all can be a significant hybrid of experiment and theory. In order to avoid the appearance of there being anything strange or paradoxical about a practice that straddles the terrain between the theoretical and the experimental, we need to recognize that while simulation is, in the general sense, a form of what we once naively called "theorizing," it is the kind of "theorizing" that has only recently begun to attract philosophical attention—the construction of local, representative models.

In the end, the point is simply this: philosophers like Hacking, Cartwright, and Giere[4] have afforded us the insight that we can have good, reliable theoretical knowledge in a particular domain and still have a lot of difficult, creative work to do in building local models under that domain. This insight is crucial if we are going to understand what goes on in simulation.

The Experimental Qualities of Simulation

Hacking, Cartwright, Giere, Morrison and Morgan, and others have shown how models can function semiautonomously from theory. What has perhaps been lacking in their analysis is an understanding of where semiautonomous models get their credentials. There is an enormous and controversial philosophical literature on how theories get credentialed. But even when it is established that the theory for a given domain is credible and reliable, how do we come to the conclusion that the local, semiautonomous models of the phenomena in that domain are reliable? As we have seen, it is not simply a matter of the model's fidelity to theory, since, in simulation, model construction often involves steps that go beyond, or even contradict, theory. It is also not simply a matter of fidelity to real-world data, since we often run simulations in order to learn more about the world than our observations will allow. Insights from the philosophy of experiment may help with these questions.

4. See, for example, Giere 1999.

A starting point for this project is to point out that simulationists and experimenters both need to engage in error management. In simulations, errors can arise as a result of transforming continuous equations into discrete ones and of transforming a mathematical structure into a computational one. All discretization techniques present the possibility that roundoff errors or instabilities may create undetected artifacts in simulation results. At a deeper level, any modeling assumption that goes into the creation of a simulation algorithm can have unintended consequences. Developing an appreciation for what sorts of errors are likely to emerge under what circumstances is an important part of the craft of both simulationists and experimenters. Precision, accuracy, error analysis, and calibration are concepts that we typically associate with experimentation and not with theorizing, but they are also very much a part of the vocabulary of the simulationist. There is indeed a great deal of similarity and analogy between the actual techniques that experimenter and simulationist each use to manage uncertainty.

In the last chapter I compared the project of the epistemology of simulation to Allan Franklin's work on the epistemology of experiment, in which he outlines a list of commonsense techniques that experimenters use to augment our reasonable belief in the results of their work. In fact, we can push the point a bit further. It is a fairly straightforward exercise to go through this list and see that many, if not all, of these techniques apply directly or by analogy to the sanctioning of simulation results.[5]

I will give a small sample here. First, Franklin argues that confidence in a piece of experimental apparatus can increase if we can use the apparatus to produce results in situations where we know what to expect, and the apparatus produces the expected results. Microscopists, for example, place known objects under their microscopes to determine if they will see what they expect. Simulationists employ similar strategies. They try to show that the simulation is capable of producing output that matches known analytic solutions to theoretically principled equations, or that the simulation can produce results that match real-world data from experiments or observations. Second, Franklin points to the strategy of showing that the apparatus will respond as expected when we intervene on the system that the apparatus is in contact with. If we double the size of an object under a microscope, we expect a properly functioning microscope to reveal an image twice as large. Similarly, simulationists will often vary the parameters and initial conditions of their models to see

---

5. See Winsberg 1999b or Weissert 1997, 122–24, for more details.

whether the simulation responds as expected. If it does, this builds confidence in the techniques being used. Another of Franklin's strategies is to show that an experimental result can be replicated with a different type of apparatus. If the optical microscope produces the same images as the electron microscope, we gain confidence in the reliability of both instruments. Simulationists also try to show that independently constructed simulation algorithms produce the same results. Depending, of course, on how different the algorithms are and on much they do or do not rely on the same sort of assumptions, this can increase confidence in the algorithms if they produce similar results. Finally, according to Franklin, experimenters try to show that the results they observe are consistent with their best theories of the systems they are studying. When experimental particle physicists uncover theoretically predicted results, we gain confidence in their methods. The situation is a bit more complicated for simulationists; since they often *use* the best theories of the system they study to construct their algorithms. But an analogous strategy is available if there are, in addition to the theoretical principles that guide the construction of the simulation, also more phenomenological laws (such as perhaps a scaling law) against which the simulation results can be compared.

There are other lessons from the philosophy of experiment that we can bring to bear here. In particular, I would like to examine the claim—found in the writings of Hacking and in Galison (1997)—that various experimental techniques and instruments develop a tradition that gives them their own internal stability, or, put most provocatively, that "experiments have a life of their own" (Hacking 1988).

A Life of Their Own

"I [once] wrote that experiments have a life of their own. I intended partly to convey the fact that experiments are organic, develop, change, and yet retain a certain long-term development which makes us talk about repeating and replicating experiments. . . . I think of experiments as having a life: maturing, evolving, adapting, being not only recycled, but quite literally, being retooled" (Hacking 1992, 307). This passage comes from a piece in which Hacking explains what it means for experiments to have "lives of their own." There are two related claims that we find being defended under that slogan by Hacking and by Galison.

The first claim is that experiments evolve and mature. Rather than being exactly replicated, experiments and instruments get adjusted to adapt to new circumstances and to incorporate new techniques for measuring and intervening. The second claim is that experiments and instruments have their own sets of credentials that they bring to scientific practice, and that these credentials develop over an extended period of time and become deeply tradition bound. This, I believe, is what Hacking intended when he said that experimental practice is "self-vindicating."

I want to argue here that both of these claims are also true of simulations. Simulation practices have their own lives: they evolve and mature over the course of a long period of use, and they are "retooled" as new applications demand more and more reliable and precise techniques and algorithms. We have only to think of simulations of the earth's climate, which have evolved over decades, to see that this claim is true. But simulations also, like experiments, gain their own credentials over time.

When I speak of "simulations" having their own lives, I am referring to the whole host of activities, practices, and assumptions that go into carrying out a simulation. This includes assumptions about what parameters to include or neglect, rules of thumb about how to overcome computational difficulties—what model assumptions to use, what differencing scheme to employ, what symmetries to exploit—graphical techniques for visualizing data, and techniques for comparing and calibrating simulation results to known experimental and observational data. Thus, it is probably most clear to say that simulation tasks, or projects, have lives of their own, and the various techniques that feature prominently in those lives carry with them their own credentials.

Whenever these techniques are employed successfully—that is, whenever they produce results that fit well into the web of our previously accepted data, our observations, the results of our paper-and-pencil analyses, and our physical intuitions; whenever they make successful predictions or produce engineering accomplishments—their credibility as reliable techniques or reasonable assumptions grows.

In other words, the next time simulationists build a model, the credibility of that model comes not only from the credentials supplied to it by the governing theory, but also from the antecedently established credentials of the model-building techniques developed over an extended tradition of employment. That is what I mean when I say that simulation practices have their own lives; and the techniques that figure prominently in those lives are self-vindicating. As simulation practices evolve and are retooled, the techniques they employ carry with them their own history of prior successes and accomplishments, and, when properly used, they can bring to the table independent warrant for belief in the models they are used to build. In this respect, simulation techniques, and indeed many precomputer calculational modeling techniques as well, are much like microscopes and bubble chambers as Hacking and Galison describe them (Hacking 1988; Galison 1997).

Consider, as an example, a particular computational technique now commonly known as the "piecewise parabolic method" (PPM). The PPM is an algorithm that has been shown to be well suited to simulating fluid flows that contain significant shock discontinuities, such as the example given at the beginning of this chapter. Different versions of the algorithm have been used to simulate a wide variety of physical phenomena, ranging from simple laboratory setups like the one described above to such complex astrophysical systems as supernova explosions, heat convection in red-giant stars, gas accretion disks, supersonic jets, and models of the development of the entire cosmos.

The PPM begins as a discretization of the Euler equations, but we should not think of it simply as a purely mathematical transformation of those equations. Recall that the primary difficulty in modeling supersonic fluids is in dealing with shock discontinuities. The difficulty arises because real fluids do not have discontinuities. In other words, the fundamental theories of fluid dynamics always describe continuous variables. Instead of discontinuities, there are very thin regions of very steep gradients.

In principle, these thin regions could be accurately modeled by including terms for viscosity and heat conduction in the principle equations. In practice, however, viscous momentum transport and molecular heat transport take place on extremely small length scales. No computer algorithm that is computationally tractable could ever hope to capture effects on these scales.

The earliest approach to solving this problem was suggested by von Neumann and Richtmyer (1950). Their solution was to artificially increase the coefficients of viscosity and heat conduction until the point at which the effects manifest themselves at length scales sufficiently large to be resolved on a reasonable computer grid. The flow can then be tweaked in just the right way to spread the shocks over a few grid cells.

The von Neumann-Richtmyer method represents the earliest attempt at overcoming these difficulties, and it is relatively simple and easy to describe. Other methods that have emerged over the last fifty years have become progressively more complex and elaborate. The PPM represents one of a broad class of state-of-the-art methods. It differs fundamentally from its predecessors in the following way. Most differencing methods are derived from Taylor series expansions of the terms in the differential equations. This move essentially assumes that the solution is smooth.

While so-called shock discontinuities are not truly discontinuous in the theory of fluids, it is nevertheless not a good assumption to treat them as smooth when you are going to discretize. So the PPM abandons this assumption, and instead of continuously piecing together linear solutions, it pieces together, in a discontinuous fashion, nonlinear, smooth solutions. Since superposition fails to apply to these nonlinear terms, this technique requires a special solver to compute the nonlinear interactions between the discontinuous states. The construction of this solver requires outside knowledge of the propagation and interaction of nonlinear waves (Winkler et al. 1987).

Exactly how all of these methods are achieved is a subject that is more than a little arcane, and the details are not that important here. What is important to note are some of the features of these methods and their development.

Perhaps most important, the presence of shocks in the flow of a system prevents any straightforward attempt to hammer the Euler equations into discrete form from being effective. Thus, there is nothing in the Euler equations or in the fundamental theory of fluids that tells you how to proceed if you are unable to capture the steep gradients in the flow that the theory predicts. The theory of fluids has been a useful guide in the development of these methods, but it has not come anywhere close to sufficing on its own, nor has it certified the end product.

Another interesting observation arises from tracing the history of these methods from von Neumann's time to the present. The history of a simulation technique is very much like the history of a scientific instrument. It begins with a relatively crude and simple technique for attacking a relatively small set of problems. Over time, the instrument or technique is called upon to attack a larger set of problems or to achieve a higher degree of accuracy. To accomplish this, the technique must be improved upon, reconfigured, and even radically revised. In each case, the knowledge relied upon to devise and sanction the tool or method can come from a wide variety of domains.

The PPM has gained recognition as a reliable method for simulating discontinuous flows over a fairly long history of use. The results of simulations making use of this technique have been evaluated and found to be reliable in a wide variety of applications. Just like scientific instruments and experimental techniques, the PPM has "matured, evolved, been adapted and not only recycled but [not quite literally] retooled" (Hacking 1992, 307). And just like instruments and techniques of their use, the trust that we place in the PPM as a reliable method has grown with every maturation, evolution, and retooling in which it has been successfully applied. Just like the microscope and the bubble chamber, the simulation practices for calculating discontinuous compressible flows, in which the PPM has figured prominently, has had its own independent history. Beginning with the von Neuman-Richtmyer method, it has matured, evolved, been adapted, recycled, and retooled. It has had a life of its own.

FOUR

A Tale of Two Methods

Imagine two physicists interested in studying the interaction of a pair of fluids at supersonic speeds. Each of them uses sophisticated technological artifacts to generate images of the flow structures that are generated as a shock wave propagates through a fluid. Each of them manipulates the equipment so as to be able to investigate their phenomenon of interest at a variety of values of basic parameters—different relative speeds, different densities of fluid, different geometrical configurations, different boundary conditions, and so on. And each of them analyzes the data and images they generate in order to try to discern fundamental patterns, scaling relations, and other features of interest in the flow.

The first physicist's equipment is a laboratory setup consisting of a tank of fluid containing simple spherical and cylindrical shapes, bubbles of gas, and a physical mechanism for causing a shock wave to propagate through the tank. The second physicist's only piece of equipment is a digital computer. Using models from the theory of fluid dynamics as a rough starting point, the second physicist builds an algorithm suitable for simulating the relevant class of flow problems and transforms that algorithm into a computer program that runs on her computer. The computer outputs data, including perhaps graphical output depicting flow patterns.

In the last chapter, we noted that simulations have epistemological features in common with experiments. But is there a fundamental difference—a difference of kind—between these two activities? And if so, how should we characterize it? How can we make precise what distinguishes activities of the kind we traditionally call "experiments" from activities of the kind we usually call "simulations." And are there, in particular, fundamental respects in which the nature of the epistemological relationship between the artifact and nature—those that involve our abilities to use the artifact to learn about nature—differs in the two cases?

One obvious difference is the special role of the computer in the second example. But there is reason to think that, at least in one respect, this difference is not entirely fundamental. It is useful here to remember that, at least on one common understanding of the notion of simulation, not all simulations are computer simulations. There is also a class of techniques for investigating nature involving something we would naturally call simulation that have nothing to do with computers. There are plenty of paradigmatic examples. Here is an amazing one: cosmologists and students of fluid dynamics can use fluids to simulate the dynamics of black holes. Here is the basic idea: We call black holes black because they have an event horizon that is, in part, a consequence of the fact that you cannot go faster than the speed of light. Many years ago, a physicist by the name of Bill Unruh noted that in certain fluids, something akin to black hole would arise if there were regions of the fluid that were moving so fast that waves would have to move faster than the speed of sound (something they cannot do) in order to escape from them (Unruh 1981). Such regions would in effect have sonic event horizons. Unruh called such a physical setup a "dumb hole" ("dumb" as in "mute") and proposed that they could be studied in order to learn things we do not know about black holes. For some time, this proposal was viewed as nothing more than a clever idea, but physicists have recently come to realize that, using Bose-Einstein condensates, they can actually build and study dumb holes in the laboratory. It is clear why we should think of such a setup as a simulation: the dumb hole simulates the black hole. Instead of finding a computer program to simulate the black holes, physicists find a fluid dynamical setup for which they believe they have a good model and for which that model has fundamental mathematical similarities to the model of the systems of interest. They observe the behavior of the fluid setup in the laboratory in order to make inferences about the black holes.

Unruh describes the idea very nicely:

[There are] similarities between the equations of motion obeyed by sound waves in a background flow, and the equations of motion of, say, a scalar field in the space-time near a black hole. Just as in the case of [a] scalar field around a black hole, the equation

of motion of the linearized sound waves is a hyperbolic equation. The background time- and space-dependent fluid flow then acts like an effective metric for the field. . . . Given the difficulties of finding small black holes to observe, [one] hope for the usefulness of such sonic analogues is in experimental tests of black hole evaporation.
(UNRUH 2002, 110–12)

Or as another author in the same volume puts it, "Even non-relativistic fluid dynamics has a Lorentzian spacetime hiding inside it" (Visser 2002, 28).

Such investigations are sometimes called "proxy experiments" (the word *proxy* having obvious connections to the idea of simulation); Unruh himself calls them "analog models," and they are sometimes just called "simulations." To distinguish them from computer simulations, I will call them "analog simulations," and hope that the reader will not mistake what I mean for computer simulations carried out on analog computers. But whether we use a term like *analog simulation,* which contains the word *simulation,* or the term *proxy experiments,* which does not, it should be clear that these are no ordinary experiments. Something like simulation seems to be involved. So perhaps there is some fundamental quality that analog simulations share with computer simulations and that jointly distinguishes them from ordinary "experiments." To repeat the question, then: is this so? And if so, what is that fundamental quality? If analog simulations count as simulations, then it cannot be the special role of the computer that is fundamental. So what, if anything, is?

Competing Intuitions

Intuitions here can tug in opposite directions. On the one hand, we are inclined to think that the first two physicists' activities—from a fundamentally metaphysical point of view—could not possibly differ from each other more. The first physicist, so this way of thinking suggests, *is generating novel empirical knowledge about fluids by manipulating an actual fluid.* The other physicist is doing no such thing. She is merely *exploring the consequences of manipulating existing knowledge*—in this case the Navier-Stokes equations—by using brute-force methods to crank out solutions to those equations that are, merely because of practical difficulties, difficult to generate by more traditional paper-and-pencil means. The following quotation encapsulates this intuition rather succinctly: "The major difference is that while in an experiment, one is controlling the actual object of interest (for example, in a chemistry experiment, the chemicals under investigation), in a simulation one is experimenting with a model rather than the phenomenon itself" (Gilbert and Troitzsch 1999, 13).

The opposite intuition fixates on the experimental qualities of simulations—even, and perhaps especially, on computer simulations—and finds no fundamental difference. Why, its proponents ask, does the second physicist often refer to what she does as conducting "numerical experiments"? Why does she call what she generates "data"? Why does simulation practice resemble experimental practice in so many obvious respects?[1] Must we dismiss all this as just loose metaphor?

Perhaps more significantly, there are troubling questions we can raise about some of the assumptions that lie behind the first intuition—assumptions that play a crucial role in painting such a stark contrast between the activities of the two physicists. The most troubling assumption, I think, is that experimenters control "the actual object of interest."

More often than not, this is simply not true. What if we were to find out that both of our original pair of physicists' primary area of interest is astrophysics? The systems that actually interest them both are the supersonic gas jets that are formed when gasses are drawn into the gravitational well of a black hole. Neither physicist, then, is actually manipulating his or her actual system of interest. Neither one is even manipulating a system of the same type, on any reasonably narrow sense of the term. Each one is manipulating something that *stands in* for the real class of systems that interest them. In one case, that stand-in is a tank of fluid. In the other, it is a digital computer. In both cases, the actual systems of interest are vastly different from the system being manipulated—in scale, in composition, and in many other respects.

This is a fairly common feature of experimental work of all kinds. Laboratory setups often differ in substantial respects from the classes of natural systems for which they are intended to speak. Think of Galileo watching the chandelier swing to learn about how all bodies fall, or Mendel manipulating his pea-plants to learn how traits are passed on from parent to offspring throughout the plant and animal kingdoms.

It might be argued, of course, that Mendel's peas and Galileo's chandelier are instances of the systems of interest, and the physicist's tank is not, but this would be begging the question. In some respects, the physicist's tank is an instance of the system of interest, since it is in fact an instance of a supersonic interaction of a pair of fluids. And few of Galileo's con-

1. See Dowling 1999 for examples.

temporaries would have thought of his chandelier as a "freely falling object." Some, conceivably, might have doubted that cultivated plants are an instance of natural heredity. The point is that what all of these examples have in common is that the object being manipulated or observed speaks for more than itself, and it takes an argument (even if that argument turns out to use, as its major premise, that one is an instance of the other) to show that it can validly do so.

So both of our physicists, and indeed almost all scientists, it turns out, rely in the end on *arguments*—either explicit or implicit—that the results obtained from manipulating their respective pieces of equipment are appropriately *probative* concerning the class of systems that interest them. The same, it seems appropriate to say, is true of almost all experimental work. And of course it is not quite correct to say (at least in the sense that Gilbert and Troitzsch intend) that anyone is "experimenting with a model." The models Gilbert and Troitszch are speaking of—the models that inspire the computer programs in computer simulations—are abstract entities,[2] and we cannot experiment with them. What simulationists manipulate is a physical entity: either a digital computer or some sort of analog device.

Of course, it could be objected that this line of reasoning does not properly distinguish between a computer program, which might be said to be the basis of a simulation, and the underlying hardware, which might be said to be accidental. On the view of such an objector, it is not a computer that a simulationist manipulates, but the computer program—an abstract entity.

It is certainly true that computer programs are multiply realizable, and what seems most salient about a computer, when it runs a simulation, is the program that it runs, and not the particular hardware that runs it. But I want to resist this objection for three reasons. The first is that, while it is tempting and indeed useful, in many contexts, to think of computation in abstract terms, there can be no real-world computation without some physical system to implement the computer. A computer program, corollarily, cannot be manipulated without manipulating the physical system that implements it—not necessarily by changing the hardware

---

2. Some commenters take the view that the "object" (in the sense I define below) of an experiment *is* a model. So, on this way of thinking, Mendel's peas were model organisms, and Galileo's pendulum is a concrete model of the freely falling object. And of course these kinds of models and many others are not abstract. I am perfectly sympathetic to this kind of talk, but notice that if we adopt it, Gilbert and Troitzsch's claim is still false—since it then becomes wrong to say that only the simulationist manipulates a model. In any case, I do not think this is the kind of model that they had in mind.

connections in the computer, but certainly by effecting some physical changes. The second reason is that if we hope to get clear on the fundamental relationships between experiment and simulation—where simulation is explicitly taken to be a kind that includes both analog and digital cases—then we are forced to take seriously the material characteristics of computation.

There is a third reason to take seriously the physical character of the computer in computer simulation. A number of philosophers have argued for continuity between simulation and experiment by explicitly conceiving of the computer in computer simulation as a physical object that is experimentally manipulated (see, for example, Humphreys 1995; Hughes 1999; Norton and Suppe 2001; and, to a limited extent, Parker 2009). To simply assert that the physical characteristics of the computer are incidental would be to beg the question against these claims. I prefer to follow all of these commenters—to take seriously the idea that computer simulationists use computers as physical stand-ins and to locate the special characteristics of simulation elsewhere.

The other problematic assumption in the line of reasoning typified by Gilbert and Troitzsch is that the physicist using the computer is not generating new knowledge but merely exploring the consequences of existing knowledge in the form of the Navier-Stokes equation. This is obviously not true in the case of analog simulations, and I hope that the arguments of chapter 2 have convinced us that it is not true in the case of computer simulations either. To think it is true is to assume that anything you learn from a computer simulation t based on a theory of fluids is somehow already "contained" in that theory. But to hold this is to exaggerate the representational power of unarticulated theory. It is a mistake to think of simulations simply as tools for unlocking hidden empirical content.

So the intuition that seems to lie behind, for example, the view articulated by Gilbert and Troitzsch, appears to be on shaky ground. Let us see if we can make the reasons for that more precise. Following the literature on this issue,[3] let us call the class of systems in which the physicists are interested (in our case, gas jets) the "target" of their investigations. And let us call the artifact that they intervene on and observe the "object" of their investigations. What Francesco Guala, Wendy Parker, and others have made perfectly clear is that both of our physicists have to establish what is sometimes called the "external validity" of the conclusions they draw from their activities.

3. See especially Guala 2002 and Parker 2009.

An experimental result is internally valid when the experimenter is genuinely learning about the actual system he or she is manipulating—when, that is, the system is not being unduly disturbed by outside interferences. An experimental result is externally valid when the information learned about the system being manipulated is relevantly probative about the class of systems that are of interest to the experimenters.[4] So each of our two physicists has the nontrivial task of establishing that what they learn about the behavior of the object of their investigations can be appropriately informative about their targets. This makes it naive to think that there is an uncomplicated sense in which the first physicist is studying nature directly, while the second one is studying only a model.

One is still bound to sense, however, that there is some kernel of truth to the first intuition. The above arguments, in other words, are unlikely to shake many from at least the suspicion, if not the conviction, that there is still something fundamentally different, something fundamentally epistemologically different, about the two kinds of activities described above. That suspicion (or conviction) is likely to be that, all of the above notwithstanding, the experimenter simply has more direct epistemic access to her target than the simulationist does. How could we try to salvage that kernel of truth?

Material versus Formal Similarity

One suggestion on how to go about this comes originally from Herbert Simon (1969), but has been made more explicit by Francesco Guala (2002). Guala is acutely aware that both experiments and simulations have objects on the one hand and targets on the other, and that, in each case, one has to argue that the object is suitable for studying the target. Despite this similarity, Guala thinks there is still a profound difference. The difference, Guala argues, is that two fundamentally different kinds of relationships can exist between an object being investigated on the one hand, and the target of that investigation on the other.

The difference, according to both Simon and Guala, is this: In an experiment, the relationship between object and target is that they share a "deep, material" similarity. In a simulation, the similarity between the object and the target systems is only abstract and formal. In the first case, "the same material causes" are at work in the object as in the target, but not in the second case. To make Guala's proposal somewhat more precise,

---

4. Parker (2009) cites Campbell 1957 as the original source of this conceptual distinction.

"experiment" and "simulation" are really two-place predicates: we should count an investigation as an "experiment" just in case the object of the investigation bears a deep material similarity to its intended target and if the same material causes are at work, and we should call it a "simulation" if the object bears only an abstract, formal similarity to its intended target.

Mary Morgan (2002; 2003) has argued for a similar view. She goes further and urges that this difference between what she calls "material" experiments and simulations is precisely what makes experiments epistemically privileged compared to simulations. The fact that the object of a simulation bears only a formal similarity to its target, according to her, makes the task of establishing a simulation's external validity—of establishing that the object is a suitable sort of entity for studying the target—that much more difficult than in the case of an experiment.

These suggestions are fairly compelling, and our example seems to support them. After all, the first physicist's apparatus, despite bearing some significant dissimilarities to an intergalactic gas jet, is still, after all, primarily composed of fluids.[5] These fluids really have different densities, and they really flow past each other at supersonic speeds. The second physicist's apparatus is made out of silicon and wire. It has none of the significant material properties of a gas jet. There is indeed an appealing sense in which the first pair shares a "material similarity" that the second pair lacks; that the same material causes (conservation of momentum, viscous forces, advection, etc.) are at work in one case but not the other. Furthermore, we have the general impression that the material similarity between object and target—the fact that the same material causes are at work; the fact that both are fluids—in the first case guarantees that there will automatically be at least *some respects* in which the results will be informative about the target. In the second case, the impression urges us, the computer can only be informative about a gas jet in virtue of being suitably programmed; the reliability of the results depends *entirely* on having chosen the right model and the right algorithm. Thinking of analog simulations, a fluid can only be informative about black holes *if* a fairly substantial assumption turns out to be correct: that there is indeed relevant formal similarity between a good model of black holes and a good model of fluids. One cannot help but be struck by this difference and find it significant, and perhaps even conclude, with Morgan, that this

5. I should note, moreover, that the claim that intergalactic gas jets are made up of "fluids" is nontrivial. See the end of the chapter for more discussion of this.

difference results in a significant disparity between the epistemic power of the simulation and that of the experiment. But despite the appeal of these suggestions, there are some obstacles that they need to overcome.

In particular, I think there are two problems with this account. Take first the claim that simulations and experiments can be distinguished by the type of similarity that obtains between the object and the target of the investigation—whether it is deep and material, or merely abstract and formal. The notion of material similarity here is too weak, and the notion of mere formal similarity too vague, to do the required work. Consider, for example, the fact that it is not uncommon, in the engineering sciences, to use simulation methods to study the behavior of systems fabricated out of silicon. The engineer wants to learn about the properties of different design possibilities for a silicon device (often a computing or a communications device), so she develops a computational model of the device and runs a simulation of its behavior on a digital computer. We examine one such example in the next chapter. Naturally, there are deep material similarities between, and some of the same material causes are at work in, the central processor of the computer and the silicon device being studied. Should we therefore conclude that the nature of this investigation is more like that of our first physicist then our second? Probably not. One problem is that, in this case, it seems clear that the *relevant* similarities are not material. This is easy to tell in this case because we know that the simulation would run equally well if the computer were made out of gallium arsenide.

The peculiarities of this example illustrate the problem rather starkly, but the problem is quite general: any two systems bear some material similarities to each other and some differences. The clear lesson of the gallium arsenide processor is that what Guala, Morgan, and Simon must have had in mind was that the *relevant* similarity between the two systems be either a material or a formal one. But this idea might be difficult to spell out in detail in a way that works. Indeed, once we put it in its proper context, the whole idea of two material entities having formal similarities becomes rather obscure. We will return to this point shortly.

The second thing that we need to recognize is that on the Simon/Guala definitions of simulation and experiment, they are both success terms. An investigation will count as an experiment only if it is successful in the sense that the relevant material similarity between object and target actually obtains, and a simulation will be successful only if the relevant formal similarity between object and target actually obtains. But this seems wrong. Surely there can be failed experiments and failed simulations. That is, surely there can be examples of experiments and simulations that fail, in the end, to be externally valid. But on the kinds of accounts offered by Guala, Morgan, and Simon, there cannot be.

There is a related worry: if experiment and simulation are success terms, then investigators may never be in a position to know if they are conducting a simulation or an experiment, since they may not know if the relevant similarity they have established is material or merely formal. Think about Galileo's famous example of dropping a mass from the mast of a moving ship. What Galileo wanted to show, of course, was that the mass would fall at the bottom of the mast and that, by extension, a mass dropped from the top of a tower would fall at the base of a tower on a rotating earth. But a critic of Galileo's argument could presumably have doubted whether the extension was legitimate. He could have doubted, as I assume some did, whether the same causes are at work when a ship is in motion as when the entire world rotates. And so according to the material similarity criterion, Galileo and his critics would have disagreed about whether the ship study was an experiment or a simulation. But this seems troubling. True, not all semantic categories need to be epistemically accessible. It does not seem to be the case, however, that we should need a God's-eye perspective to know whether something is an experiment or a simulation. It would be especially peculiar for Morgan if this were so, since she thinks that experiments are more epistemically powerful than simulations. But what good is knowing that if we can never be sure if something is an experiment or a simulation?

Simulation as Activity and as Representation

What should we conclude? One possibility is to give up on drawing a conceptual distinction between computer simulation and analog simulations on the one hand, and experiments on the other. Parker, for example, can be read as being skeptical of attempts to distinguish the kinds of activities in which our first two physicists are engaged by saying that one is doing an experiment, and the other a simulation. Instead, she argues that the distinction between the terms *simulation* and *experiment* should not be drawn in anything like the way that we have so far been inclined to draw it. In fact, I would argue that on Parker's view the two terms refer, roughly speaking, to two ontological sides of the same coin. Here is what I mean by that.

Parker defines a simulation as "a time-ordered sequence of states that serves as a representation of some other time-ordered sequence of states;

at each point in the former sequence, the simulating system's having certain properties represents the target system's having certain properties." An experiment, for Parker, is "an investigative activity that involves intervening on a system and observing how properties of interest of the system change, if at all, in light of that intervention" (Parker 2009, 4–5).

I say that, on these definitions, the terms refer to two ontological sides of the same coin because the distinction is roughly analogous to the distinction between a car and driving. For example, according to Parker's definitions, both of our physicists introduced at the beginning of this chapter are engaged in activities that involve both simulation and experiment; each term merely emphasizes different aspects of their activities. For the first physicist, the tank can *serve as a simulation* of the intergalactic gas jet, insofar as the tank undergoes a time-ordered sequence of states, and the physicist believes that the tank in each of these states is representative of what state the intergalactic gas jet would be in. What the physicist is *doing with the tank is an experiment* in so far as he intervenes on the tank in order to investigate its properties in light of that intervention. Symmetrically, for the second physicist, the computer *serves as a simulation* of the gas jet as it moves through a series of computational states, and what the physicist is *doing with the computer is an experiment*, in so far as she intervenes on the computer by putting it in a particular initial state and observes its subsequent states to learn about its properties in light of that intervention.

It is clear that on Parker's definitions there are examples of experiment that do not involve simulation (such as when I intervene on an object in order to learn only about that very object in particular). And there are also examples of simulations that are not used for experimenting (I might build an orary that simulates the solar system just to display the motions of the planets, or program a computer simulation for educational purposes). But these examples are exceptional, just as are examples of driving without a car, or using a car without driving it.

In principle, I have no problem with using these two terms in this way. The definitions are clear and useful. They are, no doubt, actually used that way in ordinary scientific parlance in some contexts.[6] Indeed, on the

---

6. It is interesting to note, however, that there are examples of computer simulation studies that do not involve "simulation" as Parker defines it. Many so-called Monte Carlo simulations, for example, produce results without doing anything like going through a sequence of states that represents the sequence of states that the target system goes through. Stop such a simulation halfway through its evolution, and the state it is in does not correspond in any way to a state of the target system. Perhaps the correct response is to deny that these are genuine simulations. But that seems far from common practice.

view of experiment that Parker and I share, this use of the term *simulation* is particularly useful. The view we share, I take it, is that, in a large class of experiments, there is some object involved that stands in for the target of interest. And Parker's definition of simulation is useful for describing, for many of those experiments, precisely what kind of "standing in" is involved. One could even refer to this view of experiment, using Parker's definition of simulation, as the "simulation account of experiment."

But I am reluctant to give up trying to draw a clear conceptual distinction between the kinds of activities that are exemplified by our two physicists. And I also think that the two sorts of activities can be usefully distinguished by using the pair of terms *experiment* and *simulation*. I think, indeed, that in most ordinary contexts, the terms are used in opposition to each other precisely to distinguish the two kinds of activities we have been discussing. And using the terms in this way helps to make sense of the fact that the word *simulation* is used to talk about computer simulations as well as analog simulations.

So I think it is worth distinguishing two rather different uses of the term *simulation*. In one sense of the word, a simulation$_R$ is a kind of representational entity. This sense of the word is covered well by Parker's definition. But in the other sense of the word, a simulation$_A$ is a kind of activity on a methodological par with, but different from, ordinary experimentation. This second sense of the word unifies computer simulation and analog simulation. And while simulation$_R$ is close to coextensive with experiment (many simulations$_R$ are experimented on, and many experiments involve representation), simulation$_A$ is meant to *distinguish* certain kinds of activities from ordinary experiments (though there may, of course, be some rare borderline or hybrid cases). With regard to simulation$_A$, there are ordinary experiments on the one hand, and there are computer simulations and analog simulations on the other.[7] Whether we want to contrast simulations with "experiments" or with "ordinary experiments"—that is, whether or not we want to think of simulations as a particular kind of experimental activity—seems to be to an issue of whether or not to award them an honorific title. And that motivation, it seems to me, is grounded in the misguided intuition that "experiments" are intrinsically epistemologically superior, which Parker is so keen to overthrow. Whenever it is convenient, I try to remember to contrast simulation with "ordinary experiment" so as not to prejudge this question. I avoid the term *material experiment* because, as should now be clear, I

7. For the remainder of the chapter, unless I specify otherwise, I will mean simulation$_A$ when I use the word *simulation*.

do not think "materiality" is fundamental to what separates simulation from ordinary experiment.

Indeed, I think it is important to clarify the distinction between what we traditionally call "experiment" (the kind of activity exemplified by the first physicist) and "simulation" (the kind exemplified by the second). Let me say first why: One of my principal interests, and one of the core topics of this book, is the epistemology of simulation. And I think that this enterprise depends crucially on our ability to sort out the epistemological respects in which simulations and experiments resemble each other and the respects in which they differ. Recall Morgan's claim that "traditional experiments" have greater epistemic power than simulations, in part because they "have greater potential to make strong inferences back to the world" (2003, 221). Parker, quite correctly I think, disputes *this* claim. She points out that in some circumstances it is substantially easier to establish the validity of a traditional experiment, but in others a computer simulation, for example, would provide arguably more reliable results.

The simple point is that the details matter. Consider once again our two physicists. If we want to know which physicist has greater potential to make strong inferences about intergalactic gas jets, we will need to know a great deal about the details of their work. How closely do the conditions in the tank mirror the conditions in intergalactic space? How much "noise" does this kind of apparatus generate? To what extent has its credibility been established through past performance? Similar questions need to be asked about the computer model. How credible is the underlying model? How crude are the approximations used in the computing scheme? How fine is the discretization grid? How many factors (viscosity, compressibility, electromagnetic forces, etc.) have been included or omitted?

So it is true that experiments are not *intrinsically* more epistemically powerful than simulations. But there may still be important epistemological differences between experiments and simulations. Indeed, I think there are. Just because experimentalists often face epistemological challenges that are just as great as those faced by simulationists, it does not follow that the *kinds* of challenges they face do not have fundamental differences. And I think they are worth spelling out.

Arguments and Background Knowledge

So let us review where we stand. It was overly simplistic to say that experiments differed from simulations in that the first investigates nature directly, while the second merely investigates a model. Both experiments and simulations involve an object and a target. And in both cases, the task of establishing the validity of using the object to make inferences about the target can be substantial and nontrivial. And in both experiments and simulations, the object of investigation is a material entity.

The material similarity proposal—the idea that the object and target of a simulation lack the kind of deep material similarity that one finds in experiments—came in response to our recognition of this state of affairs, but it was unable to overcome two obstacles. The first obstacle was that the distinction between a deep material similarity and a mere formal similarity was too vague. We considered refining the proposal to focus on the *relevant* similarities. But then the second obstacle was that it seemed wrong to define simulation and experiment in such a way that they are both success terms.

What can we do about these problems? Parker offers the helpful suggestion that the following amendments to Guala's proposal might do the trick: that simulation studies are characterized by the fact that the investigators *aim* for their objects to have *relevant* formal similarities to their targets and that ordinary experiments are characterized by the fact that the investigators *aim* for their objects to have *relevant* material similarities to their targets (2009, 4–5). Adding the word *relevant* is supposed to take care of the first obstacle, and saying that the investigators *aim* for the (material or formal) similarity instead of saying that the similarity *actually obtains* is supposed to take care of the second obstacle.

I do not think this works. I think the whole idea of formal vs. material similarity is confused, no matter how much it is tempered by "relevant," "aimed for," or whatever. First, I am puzzled by the idea of two concrete entities having objective formal similarities. Give me any two sufficiently complex entities, and I can think of ways in which they are formally identical, let alone similar. And I can think of ways in which they are formally completely different. This fact is at the heart of one of John Searle's basic complaints about computational theories of mind.[8]

Now, we can speak *loosely* and say that two things bear a formal similarity, but what we really mean is that our best formal representations of the two entities have formal similarities. Take the case of the use of fluids as analog simulators of black holes. What would it mean to say that a black hole is formally similar to a lump of fluid? This cannot be

8. See Chalmers 1996. I happen to think that Searle's specific worries about CTM have been answered by Chalmers, but not in a way that is of any help here.

62

a statement about an objective relation between two entities. The only thing it could mean is this: We believe that there are formal similarities between the mathematical structure of our best models of fluids on the one hand and our best models of highly-curved space-time manifolds on the other—and that it is this fact, rather than any material similarity between the two entities, that is being exploited by the researchers.

Properly speaking, therefore, one who claims that the researchers "aim for relevant formal similarities" between two concrete entities must really mean that they have a way in mind of modeling their target, and they have a way in mind of modeling their object, and what they hope is that on that way of thinking about the two entities, formal similarities will exist between those two models. But when you phrase it like that, what you have, in a nutshell, is the claim that simulationists aim for their objects to properly stand in for their targets—to be simulations$_R$ of their targets. But *that* is precisely what is *aimed* for in *both* cases of simulation$_A$ and ordinary experiment. At least so says the "simulation$_R$ account of experiment" to which Parker and Guala presumably subscribe.

What distinguishes simulation from ordinary experiment is what forms the *basis* for that hope. I spell out the details with more care in what follows, but roughly, it is this: In one case, we base that hope on the fact that we know how to build good models of our target systems, and in the other case, we (frequently) base that hope on the fact that the object and the target belong to the same *kind* of system—or in some cases, if you prefer, that they are materially similar. It is wrong to say that experimenters aim for their objects and targets to have material similarities. They *aim* for the one to stand in for the other, and (in many cases) they *rely* on the fact that the two belong to the same kind—and hence perhaps have material similarities—in order to argue that they are likely to achieve that aim.

It might be obvious what is coming next: If we want to characterize the difference between an experiment and a simulation, we should not focus on what objective relationship actually exists between the object of an investigation and its target, nor even on what objective relationship is being aimed for. We should focus instead on epistemological features— on how researchers *justify* their belief that the object can stand in for the target. When we do, here is what we find: What distinguishes simulations from experiments is *the character of the argument given* for the legitimacy of the inference from object to target and the *character of the background knowledge* that grounds that argument. Simulations, in particular, are legitimated by a very special kind of argument and background knowledge.

In simulation, the argument that the object can be used to stand in for the target—that their behaviors can be counted on to be relevantly similar—is supported by, or grounded in, certain aspects of model-building practice. We now need to spell out what those are.

What separates ordinary experiments from simulations are the answers to these questions: Why do the researchers believe the object can serve as a good stand-in for the target? What kind of background knowledge do they invoke, implicitly or explicitly—and does their audience need to accept—in order to argue persuasively that one can learn about the target by studying the object?

In an experiment, the argument that the object can stand in for the target can be based on a variety of possible considerations.[9] It might, for example, be based on something like the belief that the object and target are presumed to be members of the same kind or have the same material composition—this is the kernel of truth behind Simon's and Guala's proposals. What then, is the nature of the background knowledge that grounds belief in the external validity of a simulation?

The first pass at an answer would be this: In a simulation, the background knowledge that is required to argue for the external validity of the study is *trust in a model of the target systems*. But this will not work for a variety of reasons. The requirement is both too weak and too strong. It is too weak because in experiments we also need to trust in a model of the target system. For one thing, we need to have some kind of model of the target system in order to decide, for example, whether the target and object are of a kind, or have material similarities, or whatever. For another, in an experiment, when we use an object to study some target class of systems, we are holding out the object as a model of the target. So on one construal, experiments require us to have trust in model of the target systems. But the requirement is also too strong because simulationists often do not begin, by way of background knowledge, with trust in a particular model of their target systems. The hard work of simulation involves the construction of such a model. This construction makes use of other kinds of background knowledge to sanction the trustworthiness of that model. Our job here, therefore, is to clarify what kind of background knowledge that is.

So let us return to our second physicist and ask ourselves what kinds of background knowledge simulationists bring to the construction of models in computational fluid dynamics. I would argue that there are three

---

9. If there even is such an argument. Recall that in some experiments, the object is not distinct from the target, and hence no such argument is required.

kinds of background knowledge that they can bring to bear. The first is knowledge that comes from the theory of fluids. Simulationists will argue for the reliability of their simulation, in part, on the basis of the fact that the construction of their models has been guided by sound theoretical principles from the theory of fluids. Second, they will also rely on the soundness of their physical intuitions about the fluids they study. If you are simulating the flow of a river, and you are happy to treat it as if its flow is simple and laminar, for example, you might rely on the quasi-equilibrium assumption, which says that the flow of the river does not deviate too much from the steady state. How reliable this assumption will be depends entirely on how right you turn out to be about your physical intuition that the steady-state assumption is a good one. Finally, simulationists will rely, by way of background knowledge, on the soundness of the computational tricks they employ.

We could divide these into three groups. First, there are the methods by which the computational volume is discretized (e.g., finite volume, finite element, spectral, spectral element, etc.). Second, there are the methods by which one then tries—if possible—to achieve converged numerical solutions of the (continuum) nonlinear partial differential equations (i.e., using one of the aforementioned methods). Finally, there are the (often physically motivated, or sometimes outright ad hoc) models for describing physical processes that are not well described (or not at all described) by the combination of differential equations and simulation methodology. Simulations in computational fluid dynamics, for example, often rely on techniques like artificial viscosity, eddy viscosity, vorticity confinement, and others to increase the accuracy of their results. While the first of these are generally well motivated mathematically, the latter are not. And so trust, which presumably comes from a history of past successes, in these kinds of computational tricks is a third kind of background knowledge that grounds the trust in the external validity of a simulation.

Let us call these three kinds of background knowledge (and there may be other, similar ones) "principles for model building." More precisely, let us say that simulationists argue for the external validity of their simulations on the basis of their belief that they possess reliable principles for building models of the targets of their investigations.[10]

---

10. The details of the illustrative example that I use may lead some to think this account is a bit physics-centric. Perhaps it is. The extent to which basic theory plays a role is almost certainly greater in the physical sciences than in other disciplines. But I do think that the construction of most simulation models is guided by some mixture of theory, intuitive or speculative acquaintance with the system of interest (what one might, in the case of physical examples, call "physical intuition"), and

We can now, I believe, properly distinguish simulation from experiment. It is the nature of a simulation that the argument for the suitability of using the object to stand in for the target depends, by way of background knowledge, on the researchers' belief that they have reliable principles for building models (in the sense articulated above) of the very features of the target systems they are interested in learning about. Since simulations are generally used to study the dynamics of target systems, we should say that simulations are investigations in which the choice and configuration of the object of the investigation are guided and constrained by principles taken to be reliable for building *dynamical* models (abstract models that depict temporal evolution) of the target systems. And it is constitutive of simulation that the purported reliability of those principles provides the background for belief in the external validity of the investigation.

It is true that in a traditional experiment we also make use of various modeling principles in selecting/constructing the experimental system (e.g., in helping us determine what sorts of things might be confounding factors). And those modeling principles can be part of the background knowledge that sanctions the results of the experiment. But I think there are two features of the background knowledge of simulation that make it distinctive. First, the relevant model-building principles are specifically principles for building models of the *target* of the investigation. An ordinary experimentalist worries whether he can suitably control his object. For that, he needs to know how to model the object, not the target—and he is more interested in how the object is coupled to outside interferences than in its internal dynamics. Second, and perhaps more important, in simulation the reliability of the model-building principles is invoked in arguing for the *external* validity of the study—whereas when modeling principles are invoked in sanctioning an ordinary experiment, they are invoked on behalf of the *internal* validity of the study (for example, in arguing that the object has been adequately shielded, etc.).

The conceptual distinction between experiment and simulation is now clear: When an investigation fundamentally requires, by way of relevant background knowledge, possession of principles deemed reliable for building models of the target systems; and the purported reliability of those principles, such as it is, is used to justify using the object to stand in for the target; and when a belief in the adequacy of those principles

---

tried and true computational methods. In any case, I trust that even if this is not so, the expression "model-building principles" can be appropriately fleshed out in a like manner in any discipline one is inclined to study.

is used to sanction the external validity of the study, then the activity in question is a simulation. Otherwise, it is an experiment.[11] Sometimes, especially in the physical sciences, some of the model-building principles involved are guided by the theory under whose domain the behavior of the target system falls.[12]

George Platzman, speaking at a famous meteorological conference in the 1960s, made a comment that nicely reinforces this view:

I may add to this another point mentioned by Dr. Charney, a somewhat philosophical comment concerning [ordinary] experiments. I think that I agree with Dr. Charney's suggestion that machines are suitable for replacing [ordinary] experiments. But I think it is also necessary to remember that there are in general two types of physical systems which one can think of modeling. In one type of system one has a fairly good under-standing of the dynamical workings of the system involved. Under those conditions the machine modeling is not only practical but probably is more economical in a long run. . . . But there is another class of problem where we are still far from a good under-standing of the dynamical properties of the system. In that case [ordinary experiments], I think, are very effective and have a very important place in the scheme of things. (PLATZMAN, QUOTED IN SIONO 1962, 642)[13]

In fact, Guala makes similar remarks in his essay concerning the nature of the methodologies of experiment and simulation. He notes that "the knowledge needed to run a good simulation is not quite the same as [that] needed to run a good experiment" (2002, 70). This is exactly right. But I think of this not as symptomatic of the difference between (ordinary) experiment and simulation, but as constitutive of it. Guala also, I think, fails to get just right what the knowledge is that one needs. In the case of simulation, he says that one needs to know "the relationships describing the behavior" of the target systems (70). This is not a very precise claim.

11. Obviously, there is a small problem here. Since I have characterized "experiment" negatively—experiments get picked out by what they do not require—one might worry that I am letting too much in. Perhaps washing my car and whistling Dixie count as experiments. But I am just assuming, from the point of view of this chapter, that we have a pretty good preanalytic notion of what kinds of activities fall under the union of the concepts experiment and simulation, and I am only trying to characterize the difference between the two. Presumably, Parker is right that what these activities have in common is that some object is carefully set up, intervened on, and then observed in order to learn about some target.

12. These two features of computer simulations in the physical sciences—that there is an object that stands in for a target on the one hand, and that the relevant model-building principles are close to theory on the other—are responsible for motivating the intuition that computer simulation "lies somewhere between experiment and theory" that one finds so often in the literature.

13. Thanks to Wendy Parker for pointing me to this quotation. I would not go so far as to argue that Platzman is advocating precisely the same view as I am. But I do think it resonates nicely with my view.

But on an ordinary reading of it, it is clearly too much—if we had at our disposal all the relationships that described the behavior of a system, we would not need to conduct an investigation of it. As I argued above, it is too strong to say we need to have a model of the target system's behaviors. What one needs are reliable principles for building models of those behaviors. This is a very different requirement, and confusing the two is at the heart of lot of misunderstanding about simulation.[14]

The confusion is understandable since many computer simulations in physics, for example, begin with differential equations. And it is tempting to think that differential equations perfectly describe the behavior of a system. But an unsolved set of coupled, partial differential equations do no such thing. They describe how portions of the system would behave under counterfactual conditions. But until one has closed-form solutions to a set of equations, one has no description of actual behavior.

I agree with Platzman. What we need is "a fairly good understanding of the dynamical workings of the system." I would cash that out by saying that the background knowledge that simulationists need, in sum, is a set of reliable principles for building dynamical models. In the case of an analog simulation, they need reliable principles for constructing an abstract model of both the object and the target, and an argument—based in part on those principles—that the object of the investigation has been configured in such a way that the two models of these systems will have relevant similarities.

In the case of computer simulation, the object being so configured is a stored-program digital computer, and so it is configured by programming. The simulationist uses the principles deemed suitable for building models of the target to guide the construction of a computational model and uses this model to write the simulation's computer code. When a digital computer is programmed, the computer program—an abstract entity—becomes a model of the behavior of the computer qua physical system. Since the simulationist has an argument from the fact that the computer program's writing has been guided and constrained by reliable principles for building models of his target systems, he has an argument that it is also a good model of the target systems. Hence there is an argument for the external validity of the simulation that is just like the one offered for analog simulations—that both object and target have models with relevant similarities.

We can now review some of the examples from above with these criteria in mind. We can start with the more obvious ones. Take the two

14. See, for example, my discussion of Norton and Suppe's account of simulation in chapter 2.

physicists discussed at the beginning of the chapter. The first physicist studies tanks of fluid to learn about astrophysical gas jets. What makes this an experiment? What is important is the argument and its background knowledge that legitimate the study. The first physicist does not need a toolkit for building dynamical models of her target to sanction the external validity of her study. She believes the inferences she will make are legitimate because she is prepared to argue that the two systems are, in relevant respects, the same kind of system, made out of the same material, and can be expected to exhibit relevantly similar behavior. It is quite a different situation for the second physicist. He has no commitments whatsoever to the object and target being of a kind, but he must be willing to express the non-negligible hope—and to argue that such hope is well founded—that the theory of fluids, his physical intuitions about the situation of interest, and the model-building methods that have worked well in computational fluid dynamics in the past provide him with a reliable means of constructing a model of his target system. He will want to argue, in other words, that the programming of his computer has been sufficiently guided and constrained by good principles for building models of fluids to ensure that the computational model of his computer is relevantly similar to a good model of the behavior of the gas jets that interest him. And his knowledge of the theory of fluids, along with other model-building principles, plays a central role in underwriting that argument.

We can make some similar remarks about analog simulations—such as the black hole example discussed above. Here, the physicists must believe that they have good principles and methods for modeling black holes, good principles and methods for modeling fluids, and that these methods allow them to argue that the setup of the fluids they study has been guided and constrained by reliable principles for modeling black holes. They can then argue that a relevant similarity exists between a good dynamical model of the fluid, and a good dynamical model of the black holes that interest them. It is not that there is a (merely) formal similarity between black hole and fluid that makes this a simulation rather than an experiment. The relevant consideration itself is the need, by way of background knowledge, for a commitment to basic principles that guide and constrain our reasoning about these models and their similarity in the way spelled out above.

Of course, once we have carefully drawn the distinction between an experiment and a simulation, it is always possible to muddy it. Here is an example. The activity of the first physicist seems to fall squarely in the experimental camp precisely because she uses a fluid as her object of study and her target is a fluid. But there is a sense in which calling a substance a fluid is not informative. Whether or not something counts as a fluid can sometimes depend on what one wants to do with that assumption. Take, for example, the assumption that laboratory fluids are informative about astrophysical fluids. This assumption is not obvious, since astrophysical fluids are generally noncollisional. The presence of magnetic fields in astrophysical fluids, on the other hand, allows us to treat these substances as fluids despite the absence of substantial collisions. In sum, if the first physicist were using a collisional fluid in her laboratory to study a noncollisional astrophysical "fluid," it would not be entirely clear if this should count as an experiment or a simulation. We should take it to count as an experiment to the extent that we take the claim that they are both fluids to be unproblematic. But we should take it as a simulation if we think that hydromagnetic "viscosity" is only an analog of real viscosity, and, what is more important, if we think that the external validity of such a study depends on our having a dynamical model of how such viscosity arises.[15] The existence of fuzzy cases, however, does not mean that there are not paradigmatic examples of each category.

Conclusion: Epistemic Power

What about the claim that experiments are epistemically privileged relative to simulations—the claim that they "have greater potential to make strong inferences back to the world"? I think it is easy to see, following Parker, that this claim is false. A good computer simulation of the solar system—one that calculates orbits carefully from Newton's laws—will provide me with better grounds to make inferences back to the world of the planets than almost any experimental setup I can imagine because in such a case the relevant background knowledge—our ability to build good, reliable models—is virtually unassailable. How trustworthy or reliable an experiment or simulation is depends on the *quality* of the background knowledge and the skill with which it is put to use, not on which *kind* it belongs to.

But there are significant epistemological differences between simulations and experiments, and some of them might help to explain the

---

15. Interestingly, once upon a time we did not. It used to be a mystery why the extremely rare "gases" had any viscosity at all. Treating them as viscous fluids was justified entirely pragmatically. But we now have a detailed dynamical model that explains how magnetically induced instabilities give rise to something akin to viscosity.

appeal of the claim that experiments are intrinsically more epistemo-logically powerful. Take for example, the role that experiments can play in what Hempel would call "hypothesis testing." Certainly, early twentieth-century philosophers of science, both the positivists and the Popperians, overstated the importance of hypothetico-deductivism and related activities in their foundational accounts of the role of experiment in science.[16] But experiments do often play the role of providing crucial tests for theories, hypotheses, or models. This role cannot ordinarily be played by simulation,[17] since simulations, as we have noted, assume as background knowledge that we already know a great deal about how to build good models of the very features of the target system that we are interested in learning about.

What this highlights is an important epistemological facet of the dif-ference between simulation and experiment: for epistemic agents like us, experiments are epistemologically prior to simulations. In both simula-tions and experiments, you need to know something to learn something. But the knowledge you need in a simulation is always quite abstract and sophisticated, and it usually depends on things you have learned from a long history of experiment and observation. That is because we do not commit ourselves to the reliability of model-building principles unless they have been tested against experiments and observations.

One might be tempted to think that the related claim—that experi-ments are more epistemically powerful than simulations—follows from what I call the epistemological priority of experiments. But I do not think this is correct. There may have been a time in the history of science, per-haps before Newton, perhaps even earlier, when we did not have suffi-cient systematic knowledge of nature—enough of a toolkit of trustworthy model-building principles—to create a simulation that could ever be as reliable a source of knowledge as even the crudest experiment, but that time has long passed.

16. This is, in part, what the "new experimentalists" like Ian Hacking taught us with such slo-gans as "experiments have a life of their own."

17. To be more precise, there can be a role for simulation in the testing of models, but not in the same sense that I intend above. That is, we can use computer simulation to calculate what the model predicts about a particular situation, and we can compare that prediction with data from experiments and observations. But that is not the same as the role that experiments can play as we compare their results to the predictions of a model, theory, or hypothesis.

FIVE

When Theories Shake Hands

In the preceding chapters, we have devoted a fair amount of attention to the role of theory in guiding the construction of simulation models. Most of the examples we have been drawing upon, though, have been simulations whose theoretical ancestry was limited to a single domain. Some simulations, for example, are based on continuum methods, which begin with a theory that treats their objects of study as a medium described by fields and distributions. Others describe their objects of study as a collection of atoms and molecules and rely on the theoretical framework of a classical potential or as a collection of nuclei and electrons, relying on the theoretical background of a quantum Hamiltonian. And so, while we have seen that the relationship between simulation and theory is complex, our discussion of that relationship has been simplified in at least one respect.

That is because not all simulations work with the framework of a single theoretical background. Indeed, there is a class of computer simulation models that are especially important to philosophers interested in the relations between theories at different levels of description. As it turns out, certain kinds of phenomena are best studied with simulations based on theoretically hybrid models. Such models are constructed under the guidance of theories from a variety of domains. These so-called *parallel* multiscale simulation models are cobbled together using the resources of quantum mechanics, classical molecular dynamics, and the linear-elastic theory of solids.

I want to argue in this chapter that a close look at such simulation methods can offer novel insights into the kinds of relationships that exist between theories at different levels of description. Philosophers of science often assume that the interesting relationships that exist between theories at different levels are essentially mereological. But parallel multiscale models are precisely the kinds of hybrid models in which the various scales are not linked by mereology but are instead sewn together using specially constructed algorithms to mediate between otherwise incompatible frameworks.

So I need to offer an account of how it is possible for these different theoretical frameworks, which provide essentially incompatible descriptions of the underlying structure of the systems they describe, to be sewn together. One particularly interesting thing about the process is that it sometimes relies on model features that I argue should be properly understood as fictions. *Fictions,* according to the account I offer, are representations that are not concerned with truth or any of its philosophical cousins (approximate truth, empirical adequacy, etc.). This pragmatic definition of a fiction places much weaker demands on what constitutes a nonfictional representation than one traditionally encounters. Still, I argue that fictions can play an important role in science: they can help us to sew together inconsistent model-building frameworks and to extend those frameworks beyond their traditional limits.

Hybrid Models at the Nanoscale

One of the places where we find the most interesting uses of hybrid simulation models is in the so-called nanosciences. Nanoscience, intuitively, is the study of phenomena and structures, and the construction of devices, at a novel scale of description: somewhere between the strictly atomic and the macroscopic levels. Theoretical methods in nanoscience, therefore, often have to draw on theoretical resources from more than one level of description.

Take, for example, the field of nanomechanics, which is the study of solid-state materials that are too large to be manageably described at the atomic level and too small to be studied using the laws of continuum mechanics. As it turns out, one of the methods of studying these nanosized samples of solid-state materials is to run simulations of them based on hybrid models constructed out of theories from a variety of levels (Nakano et al. 2001). Such models bear interestingly novel relationships to their theoretical ancestors. So a close look at simulation methods in the nanosciences could offer novel insights into the kinds of relationships that exist between different theories (at different levels of description) and between theories and their models.

For an example of a simulation model likely to stimulate such insights, we need look no further than the so-called parallel multiscale methods of simulation. These methods were developed by a group of researchers interested in studying the mechanical properties (reactions to stress, strain, and temperature) of intermediate-sized solid-state materials. The particular case that I examine below, developed by Farid Abraham and a group of his colleagues, is a pioneering example of this method.[1] What makes the modeling technique "multiscale" is that it couples together the effects described by three different levels of description: quantum mechanics, molecular dynamics, and continuum mechanics.

Multiscale Modeling

Modelers of nanoscale solids need to use these multiscale methods—the coupling together of different levels of description—because each theoretical framework is inadequate on its own at the scale in question. The traditional theoretical framework for studying the mechanical behavior of solids is continuum mechanics (CM). CM provides a good description of the mechanics of macroscopic solids close to equilibrium. But the theory breaks down under certain conditions. CM, particularly the flavor of CM that is most computationally tractable—linear-elastic theory—is no good when the dynamics of the system are too far from equilibrium. This is because linear-elastic theory assumes that materials are homogeneous even at the smallest scales, but we know this is far from the truth. It is an idealization. When modeling large samples of material, this idealization works because the sample is large enough that one can effectively average over the inhomogeneities. Linear-elastic theory is in effect a statistical theory. But as we get below the micron scale, the fine-grained structure begins to matter more. When the solid of interest becomes smaller than approximately one micron in diameter, this "averaging" fails to be adequate. Small local variations from mean structure, such as material

---

1. Good review literature on parallel multiscale simulation methods for nanomechanics appears in Abraham et al. 1998; Broughton et al. 1999; and Rudd and Broughton 2000.

decohesions—an actual tearing of the material—and thermal fluctuations begin to play a significant role in the system. In sum, CM cannot be the sole theoretical foundation of nanomechanics—it is inadequate for studying solids smaller than one micrometer in size (Rudd and Broughton 2000).

The ideal theoretical framework for studying the dynamics of solids far from equilibrium is classical molecular dynamics (MD). This is the level at which thermal fluctuations and material decohesions are most naturally described. But computational issues constrain MD simulations to about 10^7–10^8 molecules. In linear dimensions, this corresponds to a constraint of only about fifty nanometers.

So MD methods are too computationally expensive, and CM methods are insufficiently accurate for studying solids that are on the order of one micron in diameter. On the other hand, parts of the solid in which the far-from-equilibrium dynamics take place are usually confined to regions small enough for MD methods. So the idea behind multiscale methods is that a division of labor might be possible—use MD to model the regions where the real action is, and use CM for the surrounding regions, where things remain close enough to equilibrium for CM to be effective.

There is a further complication. The propagation of cracks through a solid involves the breaking of chemical bonds. But the breaking of bonds involves the fundamental electronic structure of atomic interaction. So methods from MD (which uses a classical model of the energetic interaction between atoms) are unreliable right near the tip of a propagating crack. Building a good model of bond-breaking in crack propagation requires a quantum mechanical (QM) approach. Of course, QM modeling methods are orders of magnitude more computationally expensive than MD methods. In practice, these modeling methods cannot model more than two hundred and fifty atoms at a time.

The upshot is that it takes three separate theoretical frameworks to model the mechanics of crack propagation in solid structures on the order of one micron in size. Multiscale models couple together the three theories by dividing the material to be simulated into three roughly concentric spatial regions. At the center is a very small region of atoms surrounding a crack tip, modeled by the methods of computational QM. In this region, bonds are broken and distorted as the crack tip propagates through the solid. Surrounding this small region is a larger region of atoms modeled by classical MD. In that region, material dislocations evolve and move, and thermal fluctuations play an important role in the dynamics. The far-from-equilibrium dynamics of the MD region is driven by the energetics of the breaking bonds in the inner region. In the outer

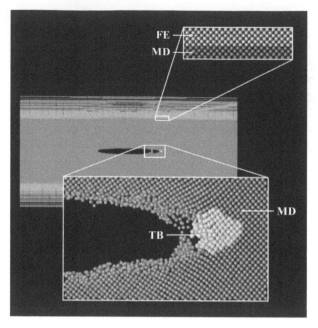

5.1 The three regions of simulation. At the center of action, where the crack propagates, the system is simulated using a tight-binding (TB) algorithm based on quantum mechanics. In the surrounding area, the far-from-equilibrium disturbances caused by the crack are simulated using molecular dynamics (MD). The elastic waves in the outlying areas are simulated using a finite-element (FE) method. Reprinted with permission from Abraham et al. 1998. Copyright 1998, American Institute of Physics.

region, elastic energy in dissipated smoothly and close to equilibrium on length scales that are well modeled by the linear-elastic, continuum mechanical domain. In turn, it is the stresses and strains applied on the longest scales that drive the propagation of the cracks on the shortest scales (see figure 5.1).

It is the interactions between the effects on these different scales that lead students of these phenomena to describe them as "*inherently* multiscale*" (Broughton et al. 1999, 2391). What they mean is that there is significant feedback between the three regions. All of these effects, each one of which is best understood at its own unique scale of description, are strongly coupled together. Since they all interact simultaneously, all three of the different modeling regions must be coupled together and modeled simultaneously. The fact that three different theories at three different levels of description need to be employed makes the models "multiscale." The fact that these different regions interact simultaneously, that they are strongly coupled together, means that the models must be "parallel multiscale" models.

An instructive way to think about the meaning of the phrase *parallel multiscale* is to compare two different ways of going about integrating different scales of description into one simulation. The first and more traditional method is what Abraham's group label "serial multiscale." The idea of serial multiscale is to choose a region, simulate it at the lower level of description, summarize the results into a set of parameters digestible by the higher-level description, and then pass those results up to a simulation at the higher level.[2]

But serial multiscale methods are not effective when the different scales are strongly coupled together. There is a large class of problems for which the physics is inherently multiscale; that is, the different scales interact strongly to produce the observed behavior. It is necessary to know what is happening simultaneously in each region since each one is strongly coupled to the others (Broughton et al. 1999, 2391).

What seems to be required for simulating an inherently multiscale problem is an approach that simulates each region simultaneously, at its appropriate level of description, and then allows each modeling domain to continuously pass relevant information back and forth between regions—in effect, a model that seamlessly combines all three theoretical approaches. This kind of method is referred to as parallel multiscale modeling or as "concurrent coupling of length scales." What allows the integration of the three theories to be seamless is that they overlap at the boundaries between the pairs of regions. At these boundary regions, the different theories "shake hands" with each other. The regions are called "handshaking regions," and they are governed by "handshaking algorithms." I discuss how this works in more detail in the next section.

The use of these handshaking algorithms is one of the things that make parallel multiscale models interesting. Parallel multiscale modeling, in particular, appears to be a new way to think about the relationship

---

2. Note that the essential idea behind parallel multiscale simulation is not entirely new. It underlies virtually all of the early attempts at subgrid modeling, including the one I discuss at length in chapter 6, namely, the original proposal for computing an artificial viscosity (i.e., using global information about the gradients of the velocity to compute a local quantity, the artificial viscosity). One fundamental difference between that kind of parallel multiscale modeling and the kind discussed here, however, is that artificial viscosity does not come from a basic theory. Indeed most subgrid modeling schemes have this relatively ad hoc character. But in the case discussed here, the schemes being used at the smallest levels actually come from the most fundamental physics. Thus, subgrid schemes like artificial viscosity do not really invoke relations between theories at different levels of description as these methods do.

between different levels of description in physics, chemistry, and engineering. Typically, after all, we tend to think about relationships between levels of description in mereological terms: a higher level of description relates to a lower level of description more or less in the way that the entities discussed in the higher level are composed of the entities found in the lower level. That kind of relationship, one grounded in mereology, accords well with the relationship that different levels of models bear to each other in what the Abraham group calls serial multiscale modeling. But parallel multiscale models appear to be a different way of structuring the relationship between different levels of description in physics and chemistry.

I would like to offer a bit more detail about how these models are put together and, in particular, to say a bit more about how the handshaking algorithms work—in effect, to illustrate how one seamless model can integrate more than one level of description. To do this, though, I must first say a bit more about how each separate modeling level works.

Three Theoretical Approaches

Continuum Mechanics (Linear-Elastic Theory)

The basic theoretical background for the model of the largest scale regions is linear-elastic theory, which relates, in linear fashion, stress (a measure of the quantity of force on a point in the solid) with strain (a measure of the degree to which the solid is deformed from equilibrium at a point). Linear-elastic theory, combined with a set of experimentally determined parameters for the specific material under study, enables you to calculate the potential energy stored in a solid as a function of its local deformations. Since linear-elastic theory is continuous, it must be discretized in order to be used in a computational model. This is done using a "finite-element" method. This technique involves a "mesh" made up of points that effectively tile the entire modeling region with tetrahedral or some other tiling. The size of each tetrahedron can vary across the material being simulated according to how much detail is needed in that area. Each mesh point is associated with a certain amount of displacement—the strain field. At each time step, the total energy of the system is calculated by "integrating" over each tetrahedron. The gradient of this energy function is used to calculate the acceleration of each grid point, which is in turn used to calculate its position for the next time step. And so on.

Molecular Dynamics

In the medium-scale regions, the basic theoretical background is a classical theory of interatomic forces. The model begins with a lattice of atoms. The forces between the atoms come from a classical potential energy function for silicon proposed by Stillinger and Weber (1985). The Stillinger-Weber potential is much like the Leonard-Jones potential in that its primary component comes from the energetic interaction of nearest-neighbor pairs. But the Stillinger-Weber potential also adds a component to the energy function from every triplet of atoms, proportional to the degree to which the angle formed by each triplet deviates from its equilibrium value. Just as in the finite-element case, forces are derived from the gradient of the energy function, which are in turn used to update the position of each atom at each time step.

Quantum Mechanics

The very smallest regions of the solid are modeled as a set of atoms whose energetic interaction is governed not by classical forces but by a quantum Hamiltonian. The quantum mechanical model they use is based on a semi-empirical method from computation quantum chemistry known as the "tight-binding" method. It begins with the Born-Oppenheimer approximation, which separates electron motion and nuclear motion, and treats the nuclei as basically fixed particles as far the electronic part of the problem is concerned. The next approximation is to treat each electron as basically separate from the others and confined to its own orbital. The semi-empirical part of the method uses empirical values for the matrix elements in the Hamiltonian of these orbitals. For example, the model system that Abraham's group has focused on is solid-state silicon. Thus the values used for the matrix elements come from a standard reference table for silicon—derived from experiment. Again, once a Hamiltonian can be written down for the whole system, the motions of the nuclei can be calculated from step to step.

Handshaking between Theories

Clearly, these three different modeling methods embody mutually inconsistent frameworks. They each offer fundamentally different descriptions of matter, and they each offer fundamentally different mathematical functions describing the energetic interactions among the entities they describe. The overarching theme is that a single Hamiltonian is defined for the entire system (Broughton et al. 1999, 2393).

The key to building a single coherent model out of these three regions is to find the right handshaking algorithm to pass the information about what is going on in one region that will affect a neighboring region into that neighbor. One of the difficulties that beset earlier attempts to exchange information between different regions in multiscale models was that they failed, badly, to conserve energy. The key to Abraham's success in avoiding this problem is that his group constructs the handshaking algorithms in such as way as to define a single expression for energy for the whole system. The expression is a function of the positions of the various "entities" in their respective domains, whether they be mesh elements, classical atoms, or the atomic nuclei in the quantum mechanical region.

The best way to think of Abraham's handshaking algorithms then, is as an expression that defines the energetic interactions between, for example, the matter in the continuum mechanical region with the matter in the molecular dynamical regions. But this is a strange idea indeed—to define the energetic interactions between regions—since the salient property possessed by the matter in one region is a (strain) field value, while in the other it is the position of a constituent particle, and in the third it is an electron cloud configuration. To understand how this is possible, we have to simply look at the details in each case.

Handshaking between CM and MD

To understand the CM/MD handshaking algorithm, first envision a plane separating the two regions. Next, recall that in the finite-element method of simulating linear-elastic theory, the material to be simulated is covered in a mesh that divides it up into tetrahedral regions. One of the original strengths of the finite-element method is that the finite-element (FE) mesh can be varied in size to suit the simulation's needs, allowing the simulationists to vary how fine or coarse the computational grid is in different locations. When the finite-element method is being used in a multiscale model, this feature of the FE mesh becomes especially useful. The first step in defining the handshake region is to ensure that as you approach the plane separating the two domains from the finite-element side, the mesh elements of the FE domain are made to coincide with the atoms of the MD domain. (Farther away from the plane, the mesh will typically get much coarser.)

FE Region   **MD Region**

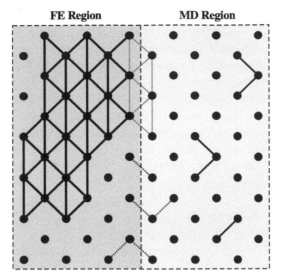

5.2   The dots in the handshake region play the role of mesh elements when they look to the left and of atoms when they look to the right. Reprinted with permission from Abraham et al. 1998. Copyright 1998, American Institute of Physics.

The next step is to calculate the energy of the handshake region. This is the region between the last mesh point on one side and the first atom on the other. The technique that Abraham's group uses is essentially to calculate this energy twice: once from the perspective of FE, and once from the perspective of MD, and then average the two. Doing the first of these involves pretending that the atoms in the first row are actually mesh elements; doing the second involves the opposite—pretending that the mesh elements in the last row are atoms (see figure 5.2).

Suppose, for, example that there is an atom on the MD side of the border. It looks over the border and sees a mesh point. For the purpose of the handshaking algorithm, we treat that mesh point as an atom, calculate the energetic interaction according to the Stillinger-Weber potential, and we divide it by two (remember, we are going to be averaging together the two energetics). We do this for every atom/mesh-point pair that spans the border. Since the Stillinger-Weber potential also involves triples, we do the same thing for every triple that spans the border (again dividing by two). This is one-half of the "handshaking Hamiltonian." The other half comes from the continuum dynamics' energetics. Whenever a mesh point on the CM side of the border looks over and sees an atom, it pretends that atom is a mesh point. Thus, from that imaginary point of view, there are complete tetrahedra that span the border (some of whose vertices are mesh points that are "really" atoms). Treating the position of that atom as a mesh-point position, the algorithm can calculate the strain in that tetrahedron and integrate over the energy stored in the tetrahedron. Again, since we are averaging together two Hamiltonians, we divide that energy by two.

We now have a seamless expression for the energy stored in the entire region made up of both the continuous solid and the classical atoms. The gradient of this energy function dictates how both the atoms and the mesh points will move from step to step. In this way, the happenings in the CM region are automatically communicated to the molecular dynamics region, and vice versa.

Handshaking between MD and QM

The general approach for the handshaking algorithm between the quantum region and the molecular dynamics region is similar: the idea is to create a single Hamiltonian that seamlessly spans the union of the two regions. But in this case, there is an added complication. The difficulty is that the tight-binding algorithm does not calculate the energy locally. That is, it does not apportion a value for the energy for each interatomic bond; it calculates energy on a global basis. Thus, there is no straightforward way for the handshaking algorithm between the quantum and MD regions to calculate an isolated quantum mechanical value for the energetic interaction between an outermost quantum atom and a neighboring innermost MD atom. But it needs to do this in order to average it with the MD value for that energy.

The solution that Abraham and his group have developed to this problem is to employ a trick that allows the algorithm to localize that QM value for the energy. The trick is to employ the convention that at the edge of the QM region, each "dangling bond" is "tied off" with an artificial univalent atom. To do this, each atom location that lies at the edge of the QM region is assigned an atom with a hybrid set of electronic properties (see figure 5.3). In the case of silicon, what is needed is something like a silicon atom with one valence electron. These atoms, called "silogens," have some of the properties of silicon and some of the properties of hydrogen. They produce a bonding energy with other silicon atoms that is equal to the usual Si-Si bond energy, but they are univalent like a hydrogen atom. This is made possible by the fact that the method is semi-empirical, and so fictitious values for matrix elements can simply be assigned at will. The result is that the silogen atoms do not energetically interact with their silogen neighbors, which means that the algorithm

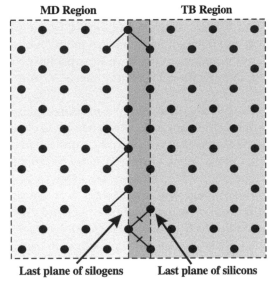

5.3 The dots in the middle region act as classical molecules in the contributions to the MD Hamiltonian and as either silicon or "silogen" atoms in their contributions to the QM Hamiltonian. Reprinted with permission from Abraham et al. 1998. Copyright 1998, American Institute of Physics.

can localize their quantum mechanical energetic contributions. Finally, once the problem of localization is solved, the algorithm can assign an energy between atoms that span the regional threshold that is the average of the Stillinger-Weber potential and the energy from the Hamiltonian in the tight-binding approximation. Again, this creates a seamless expression for energy.

Confronting Some Philosophical Intuitions

In the next section, I suggest that there are features of these multiscale models—with their integration of different levels of description, their handshaking algorithms, and their silogens—that appear on their face to be at odds with some basic philosophical intuitions about the relationships between different theories and between theories and their models. But before I begin to draw any philosophical conclusions, I think it is important to note that this area of research—nanomechanics in general and these multiscale methods in particular—is in its relative infancy. And while Abraham and his group have had some success with their models, researchers in these areas are still facing important challenges. It is probably too early to say whether or not this particular method of simulation will turn out, in the great scheme of things, to be the right way to go about predicting and representing the behavior of "intermediate-sized" samples of solid-state materials. Hence, it is probably also too early to be drawing conclusions, methodological or otherwise, from these sorts of examples.

It might *not* be too early, however, to start thinking about what kinds of basic philosophical intuitions about science are likely to come under pressure—or to be informed in novel ways—if and when these scientific domains mature. So we might, at this stage, try to pinpoint some basic philosophical questions whose answers are likely to be influenced by this kind of work. In other words, what I want to do here is to offer some ideas about what kinds of questions philosophers are likely to be able to shed light on, prospectively, if they keep an eye on what is going on in nanoscale modeling and simulation—especially with regard to multiscale methods—and to provide a sneak preview of what we might discover as the field progresses.

One issue that has received perennial attention from philosophers of science is that of the relationship between different levels of description. Traditionally, the focus of this inquiry has been debate about whether or not, and to what extent or in what respect, laws or theories at higher levels of description are reducible to those at a lower level.

Underlying all of this debate, I believe, has been a common intuition: the basis for understanding interlevel interaction—to the extent that it is possible—is just applied mereology. In other words, to the extent that the literature in philosophy of science about levels of description has focused on whether and how one level is reducible to another, it has implicitly assumed that the only interesting possible relationships are logical ones—that is, intertheoretic relationships that flow logically from the mereological relationships between the entities posited in the two levels.[3]

But if methods that are anything like those described above become accepted as successful in nanoscale modeling, that intuition is likely to come under pressure. The reason is that parallel multiscale modeling methods are forced to develop relationships between the different levels that are perhaps suggested, but certainly not logically determined, by their mereology. Rather, developing the appropriate relationships, in Abraham's words, "requires physical insight."

---

3. An important exception is the recent work of Robert Batterman (2002).

What this suggests is that there can be a substantial physics of inter-level interaction—a physics that is guided but not determined by either the theories at each level or the mereology of their respective entities. Indeed, whether or not the relationships employed by Abraham and his group will turn out to be the correct ones is an empirical/physical question and not a logical/mereological one.

Another issue that has recently begun to receive attention among philosophers of science, particularly in the work of Mathias Frisch (2004), is the importance of consistency in a set of laws. Using classical electrodynamics (CED) as an example, Frisch has challenged a common philosophical intuition about scientific theories: that the internal consistency of their laws is a necessary condition that all successful theories must satisfy. I want to make a similar point here. In this case, the example of multiscale modeling seems to put pressure on a closely related, if somewhat weaker, intuition: that an inconsistent set of laws can have no models.

In a formal setting, this claim is obviously true; indeed, it is true by definition. But rarely in scientific practice do we actually deal with models that have a clear formal relationship to the laws that inspire them. Most likely, the intuition that inconsistent laws cannot produce a coherent model in everyday scientific practice rests as much on pragmatic considerations as it does on the analogy to formal systems: how, in practice, could mutually conflicting sets of laws guide the construction of a coherent and successful model? We can start by looking at what we learn from Frisch. In CED the strategy is usually to keep the inconsistent subsets of the theory properly segregated for a given model.

The Maxwell-Lorentz equations can be used to treat two types of problems. We can appeal to the Maxwell equations to determine the fields associated with a given charge and current distribution, or we can use the Lorentz force law to calculate the motion of a charged particle in a given external electromagnetic field (Frisch 2004, 529). In other words, in most models of CED, each respective model draws from only one of the two mutually inconsistent "sides" of the theory. This technique works for most applications, but there are exceptions where the method fails. Models of synchrotron radiation, for example, necessarily involve both mutually inconsistent parts of the theory.

There are problems, in other words, that require us to calculate the field from the charges, as well as to calculate the motion of the charges from the fields. But the solution method, even in the synchrotron case as Frisch describes it, is still a form of segregation. The segregation is temporal. You break the problem up into time steps: in one time step you use the Lorentz equations; in the next you use the Maxwell equations, and so on.

A form of segregation is employed in multiscale modeling as well, but it is forced to break down at the boundaries. Each of the three theoretical approaches is confined to its own spatial region of the system.[4] But the fact that there are significant simultaneous and back-and-forth interactions between the physics in each of these regions means that the strategy of segregation cannot be entirely effective. Parallel multiscale methods require the modeler to apply, in the handshaking region, two different sets of laws. The laws in Abraham's model, moreover, are each pair-wise inconsistent. They offer conflicting descriptions of matter and conflicting accounts of the energetic interactions between the constituents of that matter. But the construction of the model in the handshaking regions is guided by both members of the pair. When you include the handshaking regions, parallel multiscale models are—all at once—*models of an inconsistent set of laws.*

The methods developed by these researchers for overcoming these inconsistencies (the handshaking algorithms) may or may not turn out to be too crude to provide a reliable modeling approach. But by paying close attention to developments in the field of nanoscale modeling, a field in which the models are almost certainly going to be required to involve hybrids of classical, quantum, and continuum mechanics, philosophers are likely to learn a great deal about how inconsistencies are managed. In the process, we will be forced to develop richer accounts of the relationships between theories and their models—richer accounts, in any case, than the one suggested by the analogy to formal systems.

Finally, these modeling techniques raise a third issue, a variation on a perennial theme in the philosophy of science: How do models differ from ideal descriptions? In particular, what role can falsehoods play in model building?

It has been widely recognized that many successful scientific models do not represent exactly. A simple example: the model of a simple harmonic oscillator can quite successfully predict the behavior of many real physical systems, but it provides at best only an approximately accurate representation of those systems. Nevertheless, many philosophers

4. I should note that multiscale modeling does not necessarily involve coupling spatially disjoint regions; it is also possible for a multiscale method to focus on one region, within which the system under study is modeled on more than one level of detail (viz., using both MD and continuum descriptions on the very same domain). There are many possible reasons for doing so—one being that a problem might call for knowing both the mean behavior (modeled adequately by a continuum approach) and the behavior of molecular fluctuations (modeled by MD) in the same region.

hold to the intuition that successful models differ from ideal descriptions primary in that they include idealizations and approximations. Ronald Laymon has made this intuition more precise with the idea of "piecewise improvability" (Laymon 1985). The idea is that while many empirically successful models deviate from ideal description, a small improvement in the model (that is, a move that brings it closer to an ideal description) should always result in a small improvement in its empirical accuracy.

But what about the inclusion of silogen atoms in multiscale models of silicon? Here, piecewise improvability seems to fail. If we make the model "more realistic" by putting in more accurate values for the matrix elements at the periphery of the QM region, then the resulting calculation of the energetic interactions in the handshake region will become less accurate, not more accurate, and the overall simulation will fail to represent accurately at all.

This would suggest that we can add certain sorts of elements to models that are different in kind from ordinary idealizations, approximations, and simplifications. I suggest that we should call such modeling elements, such as the silogen atom, *fictions*.

Fictions in Science

To a first approximation, fictions are representations that do not concern themselves with truth. Science, to be sure, is full of representations. But the representations offered to us by science, we are inclined to think, are supposed to aim at truth (or at least one of its cousins: approximate truth, empirical adequacy, reliability). If the proper and immediate object of fictions is contrary to the aims of science, what role could there be for fictions in science?

I want to use the silogen example to argue for at least one important role for fictions in science, especially in computer simulation. Fictions, I contend, are sometimes needed for extending the useful scope of theories and model-building frameworks beyond the limits of their traditional domains of application. One especially interesting way in which they do this is by allowing model builders to sew together incompatible theories and apply them in contexts in which neither theory by itself will do the job.

I have already mentioned piecewise improvability as one way in which the use of the silogen atom differs from other sorts of modeling assumptions like idealizations and approximations. But I would like to clarify more precisely what I take it to mean for a representation in science to be a fiction. My view differs substantially from those of many others who discuss the role of fictions in science. The history of discussions of fictions in the philosophy of science goes back at least to Hans Vaihinger's famous book, *The Philosophy of 'As If'*. On Vaihinger's view, science is full of fictions.[5] I believe this view is shared by many who discuss fictions in science, including many contributors to a recent volume on the subject (Suarez 2009). Vaihinger asserts that any representation that contradicts reality is a fiction or, at least, a semifiction. (A full fiction, according to Vaihinger, is something that contradicts itself.) Accordingly, most ordinary models in science are a kind of fiction.

"Fictional," however, is not the same thing as "inexact" or "not exactly truthful." Not everything, I would argue, that diverges from reality, or from our best accounts of reality, is a fiction. Many of the books to be found in the nonfiction section of your local bookstore contain claims that are inexact or even false. But we do not, in response, ask to have them reshelved in the fiction section. An article in this morning's newspaper might make a claim, "The green zone is a 10 km² circular area in the center of Baghdad," that is best seen as an idealization. And though I live at the end of a T-intersection, "Google maps" shows the adjacent street continuing on to run through my home. Still, none of these things are fictions.

I take as a starting point, therefore, the assumption that we ought to count as nonfictional many representations in science that fail to represent exactly; even representations that in fact contradict what our best science tells us to be the case about the world. Many of these kinds of representations are best captured by our ordinary use of the word "model." The frictionless plane, the simple pendulum, and the point particle all serve as good representations of real systems for a wide variety of purposes. All of them, at the same time, fail to represent exactly the systems they purport to represent or, for that matter, any known part of the world. They all incorporate false assumptions and idealizations.

But, contra Vaihinger, I urge that, because of their function (in ordinary contexts), we continue to call these sorts of representations "models," and resist calling them fictions. It seems to me to be simply wrong to say that ordinary models in science are not concerned with truth or any of it cousins. In sum, we do not want to get carried away. We do not want all (or almost all) of the representations in science, on maps, and in

---

5. See Hans Vaihinger, *The Philosophy of 'As If': A System of the Theoretical, Practical and Religious Fictions of Mankind*, trans. C. K. Ogden (New York: Barnes and Noble, 1968); orig. pub. in England by Routledge and Kegan Paul, 1924. My understanding of Vaihinger comes almost entirely from Fine 1993.

journalism, and so on to count as fictions. To do so risks not only giving a misleading overall picture of science ("all of science is fiction!") but also of weakening to the point of emptiness a useful dichotomy between different kinds of representations—the fictional and the nonfictional. If only exact representations are nonfictions, then even the most ardent scientific realist will have to admit that there are precious few nonfictional representations in the world.

Fictions, then, are more rare in science than are models, because most models in science aim at truth or one of its cousins. But fictions do not. It might seem then, that on this more narrow conception of what it is for a representation to be a fiction, there will probably turn out to be no fictions in science. Part of my goal, then, is to show that there are. To do this, I must define my more limited conception of what constitutes a fiction. This will involve being more clear about what it means to "aim at truth or one of its cousins."

So how should we proceed in demarcating the boundary between fictions and nonfictions? The truly salient difference between a fictional and a nonfictional representation, it seems to me, rests with the proper *function* of the representation. Indeed, I argue that we should count any representation—even one that misrepresents certain features of the world (as most models do)—as a nonfiction if we offer it *for* the sort of purpose for which we ordinarily offer nonfictional representations.

I offer, in other words, a pragmatic rather than a correspondence conception of fictionality. What, then, is the ordinary function of nonfictional representations? I suggest that, under normal circumstances, when we offer a nonfictional representation, we offer it for the purpose of being a "good enough" guide to the way some part of the world is, for a particular purpose. Here, "good enough" implies that the model is accountable to the world (in a way that fictions are not) in the context of that purpose. On my account, to hold out a representation as a nonfiction is ipso facto to offer it *for* a particular purpose and to *promise* that, for that purpose, the model "won't let you down" when it comes to offering guidance about the world for that purpose. In short, nonfictional representations promise to be *reliable* in a certain sort of way.

But not just in any way. Consider an obvious fiction: the fable of the grasshopper and the ant. The fable of the grasshopper and the ant, you may recall, is the story of the grasshopper who sings and dances all summer, while the ant toils at collecting and storing food for the coming winter. When the winter comes, the ant is well prepared, and the grasshopper, who is about to starve, begs for his charity. This is what we might call a didactic fiction. The primary function of the fable, one can assume, is to offer us important lessons about the way the world is and how we should conduct ourselves in it. For the purpose of teaching children about the importance of hard work, planning, and the dangers of living for today, it is a reasonably reliable guide to certain features of the world (or so one might think).

So why is it a fiction? What is the difference, just for example, between didactic fictions and nonfictions if both of them can serve the purpose of being reliable guides? To answer this, I think we need to consider the representational targets of representations. The fable of the grasshopper and the ant depicts a particular state of affairs. It depicts a land where grasshoppers sing and dance, insects talk, grasshoppers seek charity from ants, and so forth. If you read the fable wrong, if you read it as a nonfiction, you will think the fable is describing some part of the world—its representational target. If you read it in this way, then you will think it is meant to be a reliable guide to the way this part of the world—this little bit of countryside where the grasshopper and the ant live—is. In short, the fable is a useful guide to the way the world is in some general sense, but it is not a guide to the way its prima facie representational target is. And that is what makes it, despite its didactic function, a fiction.

Nonfictions, in other words, are not just reliable guides to the way the world is in any old way. They describe and *point to a certain part of the world* and say, "If you want to know *about that part of the world I am pointing to,* for a certain sort of purpose, I promise to help you in that respect and not let you down." The importance of this point about the prima facie representation targets of representations will become clear later.

Fictional representations, on the other hand, are not thought to be good enough guides in this way. They are offered with no *promises of a broad domain of reliability*. Unlike most models in science, fictions do not come stamped with promissory notes that say something like "In these respects and to this degree of accuracy (those required for a particular purpose), some domain of the world—the domain that I purport to point to—is like me, or will behave like me."

So in order to understand the difference between fictional and nonfictional representations, it is crucial to understand the different functions that they are intended to serve. But intended by whom? A brief note about intentionality is order. It probably sounds, from the above, as though in my view whether or not a representation counts as a fiction depends on the intention of the author of the representation. Thus, if the author *intends* for the representation to carry with it such a promissory note, then the representation is nonfictional. This is close to my view. But following Katherine Elgin (2009), I prefer to distinguish fictions from nonfictions without reference to the intention of the author. If we find, someday, the secret diaries of James Watson and Francis Crick, and these reveal to us that they intended the double helix model as a planned staircase for a country estate, this has no bearing on the proper function of the model. On my view, what a representation is for depends not on the intention of the author, but on the community's norms of correct use.

Consider the famous tapestry hanging in the Cloisters museum in New York, the *Unicorn in Captivity*. This is a nice example of a representation. Quite possibly, the artist intended the tapestry be taken as a nonfictional representation belonging to natural history. More important, his contemporaries probably believed in unicorns. Had there been museums of natural history at the time, this might have been where the tapestry would have hung. But today, the tapestry clearly belongs in a museum of art. Deciding whether the tapestry is a fiction or a nonfiction does not depend on the intention of the author; it depends on what, on the basis of the community's norms, we can correctly take its function to be—on what kind of museum the community is likely to display it in. Once upon a time, its function might have been to provide a guide to the animal kingdom. But it no longer is.

It is clear that science is full of representations that are inexact and of representations that contradict what we know or believe to be the case about the world. It is clear, that is, that science is full of models. But are there fictions in science? If a representation is not even meant to serve as a reliable guide to the way the world is, if it is a fiction, wouldn't it necessarily fall outside of the enterprise of science?

Despite the rather liberal constraints I want to impose on what counts as a nonfiction, I believe that fictions can play a variety of roles in science. And I believe the example of the silogen atom from nanomechanics illustrates one such role. In that example, the model builders have added a fiction to their model—the silogen atom—in order to extend the useful scope of the model-building frameworks they employ beyond the limits of their traditional domains of application.

Silogen atoms, it should be clear, are fictions. To see this, we need to look at their function. But we need to be careful. If we examine the overall model that drives the simulation as a whole, it is clearly nonfictional. The representational target of the Abraham model is micron-sized pieces of silicon. And the Abraham model is meant to be reliable guide to the way that such pieces of silicon behave. To endorse such a model in the relevant respect is *to promise that,* for the purpose of designing nano-electromechanical systems, *the model will not let you down.* It will be a good enough guide to the way these pieces of silicon behave—accurate to the degree and in the respects necessary for NEMS design. Though it contradicts reality in more ways than we could probably even count, Abraham's model is a nonfiction.

But within the simulation, we can identify components of the model that, prima facie, play their own local representational role. Each point in the QM region appears to represent an individual atom. Each point in the MD region, and each tetrahedron in the FE region, has an identifiable representational target. Some of these points, however—the ones that live in the QM/MD handshaking region—are special. These points, which represent their representation targets *as* silogen atoms, do not function as reliable guides to the way the way those atoms behave. Silogens are *for* tying off the energy function at the edge of the QM region. They are *for* getting the whole simulation to work properly, not for depicting the behavior of the particular atom being modeled. We are deliberately getting things wrong locally so that we get things right globally. The silogen atoms are fictional entities that "smooth over" the inconsistencies between the different model-building frameworks and extend their scope to domains where they would individually otherwise fail. In a very loose sense, we can think of them as being similar to the fable of the grasshopper and the ant. In the overall scheme of things, their grand purpose is to inform us about the world. But if we read off from them their prima facie representational targets, we are pointed to a particular domain of the world about which they make no promises to be reliable guides.

SIX

Models of Climate: Values and Uncertainties

No book on computer simulation in the sciences would be complete without some discussion of its application to the study of our planet's climate and its future. In many respects, computer models of climate are not that different from the other kinds of computer simulations we have seen in this book. They are built out of the same motley mixture of physical theory, approximation, physical intuition, and trial and error as models of stars, thunderstorms, and nano-materials. It follows then, that many of the philosophically interesting features of global climate models are common to a wide range of simulation models; those we have discussed at length in the previous chapters. But there are at least two features of climate models that make them worthy of special attention from the point of view of philosophy of science.

The first feature is that the extremely complex modularity of global climate models leads to a novel kind of epistemological *holism*. Recent efforts in the sphere of climate model *intercomparison* reveal that modern, state-of-the-art climate models are what I call "analytically impenetrable." I mean by this that it is extremely difficult, if not all but impossible, for climate modelers to disentangle the various sources of the successes and failures of their models. I argue, moreover, that the source of this holism is a kind of genera-tive entrenchment of various elements of the models. Con-sequently, climate models are, in interesting ways, products of their specific histories.

The second unique feature of climate models is associated with the fact that these simulations and the knowledge we gain from them have enormous public policy implications. This, in turn, makes computer simulations of climate promising vehicles for thinking about connections between public policy goals and knowledge assessment; about debates in the philosophy of science, in particular, that sometimes fall under the label of "science and values." Somewhat surprisingly, the analytic impenetrability of climate models has consequences for how we should think about the role of values in climate science. So we should start with a brief introduction to the topic of the role of values in science.

It is uncontroversial that scientific research, especially scientific research that has important public policy implications, involves value judgments. This is nowhere more evident than in research on the impact of carbon emissions upon global climate. The looming prospect of severe, anthropogenic climate change is forcing us to make difficult moral and political decisions. What kind of action should be taken to curb global climate change? How much should we value our own safety, comforts, and economic opportunities in comparison to those of future generations? How much scientific evidence do we need before taking action? Should this action be voluntary or legally mandated? The importance of these questions for our future is well known, although there is still much disagreement over the appropriate way to answer them.

Yet, there are some questions regarding the role of value judgments in climate research that are not so well known, even within the climate-modeling community. In particular, while it is clear that value judgments play an important role in deciding how a given area of research should lead us to *act*, it is less clear whether such value judgments should play a role in deciding what to *believe*. In other words, while value judgments clearly play a legitimate role in the realm of *practice*, do they also play a legitimate role in the realm of *theory*?

Philosophers of science are increasingly concerned with issues such as this, although, not surprisingly, different philosophers treat these issues in very different ways. The traditional view maintains, first of all, that the following two distinctions can be drawn clearly: the distinction between theory and practice, and the distinction between so-called epistemic and non-epistemic values.[1] It then maintains that only epistemic

---

1. Throughout this chapter, I employ the traditional terminology of *epistemic* and *non-epistemic* values. I recognize, however, that the terminology is in many respects problematic. As I will make clear, the terminology arose in the context of arguments to the effect that social values, moral

values play a legitimate role within the realm of theory; non-epistemic values can and should be confined to the realm of practice.[2] According to one influential interpretation, epistemic values—which include such values as simplicity, explanatory power, internal consistency, and consistency with surrounding theories—are values that are truth-conducive, in the sense that if theory T_1 exhibits a given epistemic value and theory T_2 does not, then T_1 is, ceteris paribus, more likely than T_2 to be true; non-epistemic values, conversely, are not truth-conducive, and thus should be excluded from the realm of theory (McMullin 1983).[3]

At various points during the last century, this traditional view has been called into question. In the early and mid-twentieth century, scholars such as C. West Churchman (1948; 1956), John Dewey (1929), Philip Frank (1954), Otto Neurath (1913), and Richard Rudner (1953) argued that values traditionally thought of as non-epistemic—including, in some cases, ethical and political values—play an inevitable role in the epistemic evaluation of research. Increasing numbers of contemporary philosophers of science are arguing for this same conclusion.[4] Despite the work of those who question this ideal of value-neutrality, however, the traditional view remains the dominant one.

One of my aims in this chapter, therefore, is to investigate the question of whether it is reasonable to expect climate modelers to exclude non-epistemic values from the "internal" aspects of their research—that is, from the realm of theory. I argue that it is not. We begin by discussing one of the most influential arguments for the claim that ethical values play an ineliminable role in the evaluation of scientific research, namely,

---

values, and other broadly practical considerations can and should be excluded from the epistemic evaluation of research—hence the label *non-epistemic*. If it turns out, however, that such values do play a legitimate role in the epistemic evaluation of research, then it makes little sense to call them "non-epistemic" values. Thus, the arguments presented here—in addition to the literature cited in note 4 below—provide strong reasons for abandoning the terminology of *epistemic* and *non-epistemic* values. However, given the entrenched character of this terminology, I have decided for convenience to use it here.

2. Defenders of this view include Giere 2003; Jeffrey 1956; Kitcher 2001; Koertge 2003; McMullin 1983; and Mitchell 2004.

3. It has not always been standard to allow a role for values of any kind in the appraisal of theories. One of the primary aims of the neo-positivist movement in post–World War II America was to explicate fundamental methodological concepts such as confirmation and explanation in purely formal terms (e.g., Hempel 1945; Hempel and Oppenheim 1948). Had this project succeeded, there would be no need for any values—including epistemic values—in the appraisal of theories. The project, however, did not succeed, and since the demise of neo-positivism, it has been standard to argue that values of some sort play an inevitable role in theory appraisal. For classic expressions of this view, see Kuhn 1969 and 1977; and McMullin 1983.

4. See, for example, Douglas 2000; Howard 2006; Longino 1990 and 2002; Kourany 2003a and 2003b; Solomon 2001; and Wilholt 2009.

Richard Rudner's argument from inductive risk (Rudner 1953); and follow this with a discussion of one of the most influential objections to this view, that of Richard Jeffrey (1956).[5] Jeffrey's argument is the locus classicus of the view that one can distinguish clearly between the epistemic and the practical appraisal of theories, and that the epistemic appraisal of research can and should be neutral with respect to non-epistemic value judgments. Jeffrey's argument, moreover, is followed by many contemporary philosophers of science.[6]

An investigation of climate modeling is a particularly fruitful way to test the feasibility of Jeffrey's argument, because climate modelers, together with statisticians, attempt to do something very similar to what Jeffrey recommends—namely, to assign probabilities to hypotheses concerning the effects of carbon emissions upon global climate change in a manner that is free from non-epistemic considerations.

Before we proceed to our discussion of Rudner and Jeffrey, I must make one preliminary point. There is a broad consensus within the scientific community that carbon emissions are causally related to global climate change; I am convinced that this broad consensus is justified, and it is in no way my aim to call this consensus into question. I believe this consensus has been reached objectively. As will become apparent, I do not believe that the influence of non-epistemic considerations on the estimation of uncertainties implies that climate models are unreliable. What it does imply is that, pace Jeffrey, one cannot distinguish sharply between the realms of value-neutral theory and value-laden practice. Moreover, it highlights the fact that we need to pay more attention to the areas within climate modeling in which values play an ineliminable role, to the kinds of values or practical considerations that play a role, and to the effect that these values have upon the overall performance of our models. None of this, however, should be taken as evidence for skepticism about climate change. With this preliminary note aside, we can now proceed to our discussion of Rudner and Jeffrey.

5. Rudner's argument from inductive risk is not the only argument for the claim that non-epistemic values play an ineliminable role in the evaluation of research. Another kind of argument stems from the thesis of underdetermination and has been advanced in various forms by Neurath 1913; Howard 2006; and Longino 1990 and 2002. The focus of this paper, however, is upon Jeffrey's response to Rudner's argument from inductive risk.

6. For example, Ron Giere employs a version of this objection in his criticism of Janet Kourany's proposal for a socially responsible philosophy of science (Giere 2003; Kourany 2003a and 2003b), and Sandra Mitchell employs it against the arguments of Heather Douglas (Douglas 2000 and 2004a; Mitchell 2004).

Rudner and Jeffrey on the Role of Ethical Values in Science

In an essay entitled "The Scientist *Qua* Scientist Makes Value-Judgments," Rudner argues that ethical values play an inevitable role in the epistemic appraisal of hypotheses. His argument proceeds as follows:

1. The scientist *qua* scientist accepts or rejects hypotheses.
2. No scientific hypothesis is ever confirmed with certainty. In accepting or rejecting a hypothesis, there is always the possibility of being wrong.
3. The decision to accept or reject a hypothesis depends upon whether the evidence is sufficiently strong.
4. Whether the evidence is *sufficiently* strong is "a function of the *importance*, in a typically ethical sense, of making a mistake in accepting or rejecting the hypothesis" (Rudner 1953, 2; his emphasis).
5. Therefore, the scientist *qua* scientist makes value judgments.

To illustrate this argument, Rudner considers the evaluation of the following two hypotheses: (1) a given drug, which we know contains a toxic ingredient that is lethal to human beings, is safe for human consumption because it does not contain this ingredient in dangerous quantities; (2) a given batch of machine-stamped belt buckles is not defective. Rudner argues that we require a much higher standard for accepting the former hypothesis than the latter, because the moral consequences of wrongly accepting the first hypothesis are much more serious than the second. "How sure we need to be before we accept a hypothesis will depend on how serious a mistake would be" (Rudner 1953, 2). Given that the degree of confirmation that we require in order to accept (or reject) a given hypothesis depends upon an evaluative judgment regarding potential ethical consequences, ethical considerations play an inevitable role in the appraisal of hypotheses. Note that, in drawing this conclusion, Rudner is maintaining that value judgments, *even in the ideal*, play a role in the appraisal of hypotheses. It is not only the scientist *qua* human being—that is, *qua* individual who is invariably influenced by the prejudices of her time—who makes value judgments; it is also the scientist *qua* scientist.

While Rudner does not spend much time examining the implications of this conclusion for a broader theory of science, he does discuss briefly its implications for the notion of scientific objectivity. According to a traditional interpretation, scientific objectivity requires that hypotheses be evaluated in a value-neutral fashion—or at least in a fashion that is neutral with respect to ethical values. Rudner does not deny that objectivity is and should be an ideal for scientific inquiry, but he argues that the necessarily value-laden character of hypothesis appraisal implies that this traditional interpretation of objectivity is misguided: "What seems called for . . . is nothing less than a radical reworking of the ideal of scientific objectivity" (Rudner 1953, 6). Rudner does not undertake the task of reworking this ideal; he does, however, argue that objectivity in science "lies at least in becoming precise about what value judgments are being and might have been made in a given inquiry" (6). In other words, if value judgments do play an inevitable role in scientific research, it would be in the interests of objectivity to know precisely which value judgments are playing a role, where they are playing a role, and the effect that these judgments have upon the research in question.[7]

Many, however, deny that value judgments of an ethical—or, more broadly, non-epistemic—character play an inevitable role in the epistemic appraisal of hypotheses, and one of the clearest arguments for this view was put forward by Jeffrey, in direct response to Rudner (Jeffrey 1956). Jeffrey argued that the scientist *qua* scientist does not accept or reject hypotheses but merely assigns probabilities to them: "The scientist's proper role is to provide the rational agents in the society which he represents with probabilities for the hypotheses which on the other account he simply accepts or rejects" (245). Once the scientist has assigned probabilities to hypotheses and communicated this information to "rational agents" in society, these agents then assign utilities to the possible outcomes of accepting/rejecting the hypotheses and determine on the basis of a decision-theoretic calculation whether the hypothesis in question should be acted upon.

The primary argument that Jeffrey provides for the view that the scientist *qua* scientist does *not* accept or reject hypotheses is that the acceptance or rejection of a hypothesis per se, independent of a given practical context, is incoherent. In support of this, Jeffrey considers the following two hypotheses: (1) an entire lot of vaccine is safe, and (2) an entire lot of roller skate ball bearings is safe. In certain contexts, he argues, it is legitimate to demand a higher standard for the "acceptance" of (1) than for (2)—but only when certain practical contexts are assumed, such as that the vaccine will be given to children. If we assume a different practical context, such as that the vaccine will be given to monkeys, the standards for "acceptance" of the two hypotheses might be the same. Jeffrey sum-

7. See Douglas 2004a; Fine 1998; and Longino 1990 for discussions of the compatibility of values and objectivity in science.

marizes his position by quoting approvingly the following passage from Bruno DeFinetti:

I do not deem the usual expression "to accept hypothesis H_r," to be proper. The decision does not really consist of this "acceptance" but in *the choice of a definite action A_r.* The connection between the action A_r and the hypothesis H_r may be very strong, say "the action A_r is that which we would choose if we knew that H_r was the true hypothesis." Nevertheless, this connection cannot turn into an identification."
(QUOTED IN JEFFREY 1956, 242).

Thus, when one "accepts" a hypothesis *H*, one is really choosing to act on the basis of *H*, and to do this requires that we specify a practical context in which the action is to occur. Because virtually all hypotheses could be acted upon in a multiplicity of different ways—for example, the hypothesis that a given vaccine is safe could be used as a basis for vaccinating either children or monkeys—the notion of "accepting" a hypothesis *H* per se is incoherent.

For our purposes, the primary implications of this view are (1) that one can distinguish clearly between the realms of theory and practice, or belief and action, and (2) that ethical considerations—or, in Jeffrey's terminology, considerations of utility—can be confined to the realm of practice. The scientist, Jeffrey argues, can and should remain in the realm of value-neutral theory and leave the ethical questions to "the rational agents in the society which he represents."

As stated earlier, Jeffrey's objection to Rudner's argument is still a standard objection to the view that ethical considerations play an inevitable role in the evaluation of research. For example, in response to Heather Douglas's argument that "non-epistemic values are a required part of the internal aspects of scientific reasoning" (Douglas 2000, 559), Sandra Mitchell maintains that this argument involves a "conflation of the domains of belief and action [that] confuses rather than clarifies the appropriate role of values in scientific practice" (2004, 250). Moral and political values, according to Mitchell, play a legitimate role in the *practical*, not the *epistemic*, evaluation of research.

In the remainder of the chapter, I argue that in the area of climate modeling, the Jeffreyan strategy does not succeed. In recent work, climate scientists, working alongside statisticians, have begun to attempt to do something very similar to what Jeffrey recommends; they attempt, that is, to estimate the uncertainties of various predictions made by climate models, and they attempt to do this in a manner that is free from moral, social, or any other kind of non-epistemic value. They then hand these predictions and uncertainties over to policymakers, legislators, and other representatives of the public who are charged with determining how best to act.[8] In my view, however, this clean separation of the realms of value-neutral theory and value-laden practice is not realistically attainable. I contend that one class of non-epistemic values in particular—those reflected in deciding that certain types of prediction tasks are more important than others—influence the probabilities we assign to various possible climate outcomes. In order to argue for this, we need to establish a few preliminaries regarding the way in which uncertainties in climate modeling are estimated.

Climate Modeling and Uncertainty

Conceptually, we can distinguish three sources of uncertainty regarding the predictions of complex climate models. First, there is uncertainty about the basic structure that our climate models ought to have. While the construction of climate models is guided by basic science, these models incorporate a plethora of auxiliary assumptions, approximations, and parameterizations, all of which contribute to a degree of uncertainty about the predictions of these models. We will call this type of uncertainty *structural model uncertainty*. Second, complex models involve large sets of parameters, or aspects of the model that must be quantified before we can use it to run a simulation of a climate system. We are often highly uncertain about what are the best values for many of these parameters, and hence, even if we had at our disposal a model with *ideal structure*, we would still be uncertain about the behavior of the real system we are modeling, because the same model structure will make different predictions for different values of the parameters. We will call uncertainty from this source *parameter uncertainty*. Finally, in evaluating a particular climate model, including both its structure and parameters, we compare the model's output to real data. Climate modelers, for example, often compare the outputs of their models to records of past climate. These records can come from actual meteorological observations or from proxy data—inferences about past climate drawn from such sources as tree rings and ice-core samples. Both of these sources of data, however, are prone to error, and so we are uncertain about the precise nature of the past climate. This, in turn, has consequences for our knowledge of the future climate. We will call this source of uncertainty *data uncertainty*.

8. For a clear example of this, see the so-called Stern Review (Stern 2007).

While data uncertainty is a significant source of uncertainty in climate modeling, I do not discuss this source of uncertainty here. For the purposes of this discussion, I make the crude assumption that the data against which climate models are evaluated are known with certainty. I am interested in arguing that values play an inevitable role in the estimation of uncertainties from the two other sources. How, then, do we estimate structural model uncertainty and parameter uncertainty? In both cases, there are, broadly speaking, two methods available to statisticians interested in quantifying these uncertainties. The first method uses *observable frequencies*, and the second uses *expert judgment*.

To understand the idea of using observable frequencies, consider the example of a simulation model with one parameter and several variables.[9] If we have a data set against which to benchmark the model, we could assign a weighted score to each value of the parameter based on how well it retrodicts values of the variables in the available data set. On the basis of this score, we could then assign a probability to each value of the parameter. Crudely speaking, what we are doing in an example like this is *observing* the *frequency* with which each value of the parameter is successful in replicating known data—how many of the variables does it get right? with how much accuracy? over what portion of the time history of the data set?—and then assigning this observed-frequency value to the probability of the parameter taking this value.

Of course, only in specific circumstances are frequencies probabilities. Absent other knowledge, it would be naive to think the observed frequencies in an example like the above are the actual probabilities of the values of those parameters. For one thing, we are interested in the best value of the parameter for predicting the behavior of the system for *all* times, not just the times for which we have sample data. For another, carrying out a procedure like the one above requires us to weight the relative importance of the various variables. Hence, while such frequencies can be useful guides in assigning probabilities to the values of a parameter, they are far from perfect. Some might therefore think it more sensible to adopt a broadly subjectivist approach and to assume that the best guide to these probabilities is the subjective degree of belief held by the best experts. Expert judgment, however, is not perfect either; in fact, expert

9. A parameter for a model is an input that is fixed for all time, while a variable takes a value that varies with time. A variable for a model is thus both an input for the model (the value the variable takes at some initial time) and an output (the value the variable takes at all subsequent times.) A parameter is simply an input. We do not discuss "input variable uncertainty" because it is usually assumed, in climate science, that particular choices of input variables only cause "transients" that can be eliminated from model output after the models are "spun up."

judgment surely arises, inter alia, though the process of observing the degree to which model output matches available sample data. In practice, therefore, statisticians typically use some combination of expert judgment and observable frequencies to arrive at probabilities.

Thus, if we are interested in understanding where estimates of the degree of uncertainty about future climate come from, and, in particular, if we want to know to what degree these estimates are free from, or influenced by, various values, then we need to understand at least four things. We need to understand how observable frequencies and expert judgment are each used to estimate the degree of both structural model uncertainty and parameter uncertainty.

Structural Model Uncertainty

Let us begin with the estimation of structural model uncertainty. Given that one common method of estimating the degree of structural uncertainties in the predictions of climate models is to examine the degree of variation in the predictions of the existing set of climate models,[10] the range of models that happen to be available on the market will clearly influence our estimates of the degree of uncertainty contained in them. In this section, I argue that climate modelers have emphasized certain prediction and retrodiction tasks over others (e.g., they have emphasized predictions of global mean surface temperature change over other possible predictions, such as predictions of global mean precipitation change, ice melting, sea temperatures, etc.), and that these past decisions have affected the performance of these models. I then argue that this in turn invariably affects the current estimations of uncertainties. The decision to emphasize one prediction task over another is a paradigm example of a decision that reflects non-epistemic values; that is, the decision is made not because emphasizing one prediction task over another has significant epistemic benefits, but because one set of prediction tasks is thought to be more important, in terms of its social, political, or economic consequences, than another. These value-laden decisions, I argue, invariably affect the estimation of uncertainties of climate models.

10. See, for example, IPCC 2007; or, for a more skeptical position regarding the feasibility of this endeavor, Smith 2002. The typical method is simply to take the average and standard deviations of the existing predictions and generate uncertainties from these values.

As a preliminary to this argument, it is worth noting that the size of the dispersions of predictions of different climate models depends significantly upon the choice of a prediction task. Consider, for example, the graphs shown in figure 6.1, which display predictions of temperature change and precipitation change made by different climate models.[11] For our purposes, what is important about these graphs is that the dispersion of predictions of mean precipitation change is significantly larger than the dispersion of predictions of global mean surface temperature change. The reason for this, I hypothesize, is that predictions of global mean surface temperature change have been more highly valued than other predictions; the climate-modeling community has focused its energies upon refining and tweaking models in order to predict temperature change accurately, rather than precipitation change, sea level change, or any of a variety of other prediction tasks.

One might, however, question this hypothesis; after all, temperature and pressure are highly coupled variables in all of our models. One might expect, therefore, that improvement in one should only be able to come in tandem with improvement in the other. But this is not the case. Members of model intercomparison groups have noted that "accurate simulation of one variable does not in most cases imply equally accurate simulation of another" (Gleckler, Taylor, and Doutriaux 2008, 8). One might wonder what the source of this variation is. Is there further evidence that the degree of structural model uncertainty that we get from observable frequencies (from looking at the range of available models on the market) is affected by past values regarding the importance of various prediction tasks?

To answer this question, it is helpful to introduce what one might call the "problem of attribution." Since at least the late 1980s, the climate-modeling community has been attempting to attribute the specific successes and failures of different climate models to specific components, or modules, of those models, in order to develop models that predict all relevant quantities equally well. The thought was that once one understood the sources of disagreements among different climate models, one could take steps toward eliminating those disagreements. In 1989, a major program was founded at the Lawrence Livermore National Laboratory called the Program for Climate Model Diagnosis and Intercomparison (PCMDI), which had this as its stated goal. A number of projects were undertaken under the auspices of the PCMDI, including the Atmospheric

11. Taken from IPCC 2007.

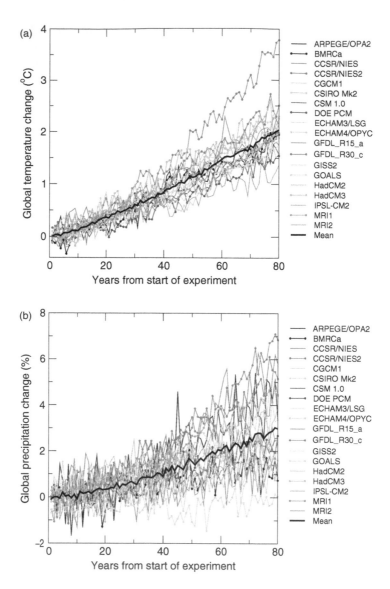

6.1 The first graph shows the predictions of a range of models for global temperature change (°C) vs. time (years). The second graph shows the predictions, for the same range of models, for global precipitation change (%) vs. time (years). Reprinted from the Intergovernmental Panel on Climate Change, *IPCC Third Assessment Report: Climate Change 2001—The Scientific Basis: Contribution of Working Group I to the Third Assessment Report of the Intergovernmental Panel on Climate Change* (Cambridge: Cambridge University Press, 2001), 537.

Intercomparison Project, which began in 1990, and the Coupled Model Intercomparison Project, which began in 1995.[12]

Despite the optimism that surrounded the founding of the PCMDI, however, the project of attribution has been unsuccessful; moreover, there are strong reasons for believing that it will continue to be unsuccessful. Because the failure of the project of attribution is highly relevant to our argument that non-epistemic values play an ineliminable role in the estimation of uncertainties, we must examine in some detail the reasons behind this failure.

One way to think about the failure of the climate intercomparison projects is in terms of a form of what philosophers call confirmation holism. Confirmation holism, as it is traditionally understood, is the thesis that a single hypothesis cannot be tested in isolation, but that such tests always depend on other theories or hypotheses. It is always this collection of theories and hypotheses *as a whole*, says the thesis, that confront the tribunal of experience. But in contrast to the way the problem of confirmation holism is typically understood in the philosophy of science, the problems faced by climate scientists are not merely logical problems, nor are they confined to the role of anything that can suitably be called auxiliary hypotheses. Rather, they are deep and entrenched problems that confront the scientist who works with models whose component parts interact in such a complex manner, and have such a complex history, that the scientist is unable to evaluate the worth of the parts in isolation.

Before returning to the issue of values, I want to argue for two claims about climate models. The first claim is about *holism*. I contend that recent efforts in the sphere of climate model *intercomparison* reveal that modern, state-of-the-art climate models are what I call "analytically impenetrable." I spell out this notion with more care in what follows, but the intuitive idea is that, as a practical matter, it has become impossible for climate scientists to *attribute*[13] the various sources of relative successes and failures to particular modeling assumptions.

The second claim is about *entrenchment*. In particular, I argue that entrenchment can be identified as one of the principal causes of holism. Here, I want to argue that climate models are, in interesting ways, products of their specific histories. Climate models are developed and adapted to specific sets of circumstances, and under specific sets of constraints,

---

12. For further details, see the PCMDI website: http://www-pcmdi.llnl.gov/about/index.php (accessed May 13, 2008).

13. A cognate of this word, *attribution*, occurs in the prominent phrase "attribution of climate change," which stands for the question whether observed climatic change is caused by humans. I do not use the word in this way here

and their histories leave indelible and sometimes inscrutable imprints on these models.

To a first approximation, we can think of the validation of a model in the following way: A model is validated when we are convinced that there is an appropriate fit between the dynamics of the model,[14] on the one hand, and the dynamics of the real-world system to be modeled, on the other. As we have seen in previous chapters, such a conception of the validation of simulation models is somewhat simplified. In particular, simulations are often used to generate predictions about phenomena in domains where data are sparse. Hence, while appropriate fit is of course what we want in a model, we want more than fit with those features of the real-world system that are immediately, observationally accessible to use. That a model is valid, therefore, is rarely established solely by comparing it to the world. As we have seen, the sanctioning of simulation models depends on a number of features in addition to fidelity of the simulation's output to known real-world data. It also depends on fidelity to theory, to accepted computation method, and a host of other factors. Here though, I want to set these complications aside, and focus, in particular, on the role of comparison with data in the validation of simulations. I also want to focus, in this chapter, on a particular facet of validation. I want, in particular, to think about situations in which models fail to be adequately validated—about situations, in other words, where the behavior of the model is known *not* to be close enough to the behavior of the world for its intended purpose.

This, after all, is the state of affairs known to obtain with regard to most global climate models. Several such models are being run by research centers worldwide. Each has its specific strengths and weaknesses in certain respects. The series of assessment reports of the Intergovernmental Panel on Climate Change (IPCC) documents how adequacy of the overall picture is thought to be produced by a synopsis of a plurality of models. In such cases, the issue of model validation is, in effect, the issue of *model improvement*. To put the central question succinctly: When a complex models fails to be adequate, is it possible to identify the various components of the model that contribute to its relative successes and failures?

It is precisely in these contexts, however, that a serious form of confirmational holism rears its ugly head. On the common understanding of this thesis, a result of the so-called Quine-Duhem problem, it is thought to have two features. First, the problem of confirmational holism is typi-

---

14. "Appropriate" in the sense that, for the intended purpose of the model, the model is close enough to the world in the intended respects and to the intended degree of accuracy.

cally associated with the idea of *auxiliary hypotheses* having to do with *observation*. Suppose, for example, that we have the hypothesis that all metal rods expand when heated. An alleged falsification of this hypothesis comes from the observation of a rod being heated and not expanding. Confirmational holism comes from the realization that such an observation's credibility depends on a sound understanding, grounded in certain theories or hypotheses, of thermometers and measuring instruments. Any apparent conflict between our original hypothesis and our data could be either the fault of the original hypothesis, or the fault of these auxiliary hypotheses—hypotheses associated with measuring instruments. Second, the problem of confirmational holism is often thought to be a *logical problem*. In other words, on a common understanding of the Quine-Duhem problem, and of confirmational holism, what we are supposed to conclude is that *logic alone* never dictates whether a single hypothesis or theory is confirmed or falsified by a collection of data. But it is usually supposed that good judgment (what Duhem called "bon sense") can decide between such rival possibilities. This is often supposed on the basis of the belief that auxiliary hypotheses used in observation can be independently tested. It is usually supposed, in other words, that the Quine-Duhem problem is a philosophical problem without actual practical implications for the working scientist.[15]

But in contrast to the conventional picture of how Quine-Duhem is supposed to operate, the holism that arises in climate modeling is wholly independent of whatever hypotheses or theories sanction the reliability of the observational base upon which validation occurs. Even when the reliability of the data against which simulation output is being compared is not in doubt—that is, even if we imagine a situation where, for example, the data concerning historical records of ice-ages against which the simulation's output will be compared are not open to question, where there is no concern about the reliability of the auxiliary hypothesis used to generate these data—there is still a serious problem of confirmational holism.

Kluges, Generative Entrenchment, and the Historical Character of Model Performance

The most sophisticated current climate models, Atmosphere-Ocean General Circulation Models (AOGCMs), are highly complex computer

---

15. For an excellent discussion of Duhem's contributions to the philosophy of science, see Darling 2002.

models that are constructed on the basis of both principled science—including fundamental partial differential equations from mechanics and thermodynamics—and trial-and-error approximations and parameterizations, and everything in between. An important feature of these models is their modular structure; each model is made up of components, or modules, and each module represents a given subsystem of the earth's climate, such as the circulation of the atmosphere, ice formation, ocean dynamics, cloud formation, the effects of vegetation, and the dynamics of aerosols. Today's climate models have their roots in the General Circulation Models of the 1950s, which described the atmosphere; over time, the complexity of these models has increased via the addition of more and more modules. The addition of these modules has allowed for better predictions of a growing range of phenomena; as already indicated, however, the predictions made by different climate models differ from one another, in some cases significantly.

An important reason for this variation in predictions is that the modules that make up a climate model are not mutually independent; rather, they interact with one other, so that the overall performance of a model depends not simply upon the representational adequacy of each individual module, but also upon the way in which the modules are coupled together. Thus, if module X is good at predicting phenomenon Y, and if module W is good at predicting phenomenon Z, it will not, at least in general, be the case that the model consisting of X and W will be good at predicting both Y and Z. This overall model might be good at predicting one or the other of these phenomena, or neither, depending upon the way in which the modules interact with one another.

Moreover, not only does the overall performance of a given climate model depend upon the interaction of its modules, but it is also the case that the specific ways in which modules interact with one another will depend upon the overall model of which they are parts. The reason for this is that the process of coupling modules together relies to a large extent upon fitting the module to the overall model—not only to its principled structure, but also to its parameterizations.[16] This, in turn relies on trial and error, piecemeal, mutual adjustments of parameters and parameterization schemes—adjustments that need to be undertaken in different ways, depending upon the details of the model and modules in question.

---

16. Parameterizations, roughly speaking, are correction factors, or attempts to correct for the fact that simulation models cannot capture effects that occur at scales below which the models are discretized. We discuss parameterizations in much greater detail in the section below entitled "Parameter Uncertainty."

In order to illuminate the nature of these adjustments, a number of commentators have employed the notion of a *kluge*, a colloquial term first employed by computer programmers to describe sections of code that were functional but unprincipled, inelegant, and ill-understood. According to Andy Clark, a kluge is "an inelegant, 'botched together' piece of program; something functional but somehow messy and unsatisfying" (Clark 1987, 278). The term describes aptly the process of coupling two modules together; while parts of the process might be principled, other parts rely to a large extent upon simply "botching" things together.

In my view, the rather messy character of module coupling has important implications for the development of climate models; it suggests that model development is a historical process that depends in important respects upon the environment in which that model develops. To see this, it is helpful to draw upon William Wimsatt's notion of "generative entrenchment" (Wimsatt 2007). According to Wimsatt, "a deeply generatively entrenched feature of a structure is one that has many other things depending on it because it has played a role in generating them" (133). Wimsatt employs this notion in order to explain the relationship between biological development and evolution; characteristics that are adaptive for one organism will not, in general, be adaptive for another, because the different organisms will, in general, have different features that are generatively entrenched. Unlike Dumbo, real elephants will never be able to fly, because they have particular features—for example, bulkiness—that have developed in specific circumstances as a result of specific environmental pressures and that make adaptations such as wings impossible. Analogously, some climate models will never be successful at predicting a particular phenomenon via a particular module, because the models have developed in specific circumstances as a result of specific "environmental" pressures—for example, pressures to emphasize certain predictive tasks over others—that make the successful inclusion of a given module impossible.

Consider the following idealized example. Climate Model *A* is a simple General Circulation Model that describes the dynamics of the atmosphere. By building upon *A*, two further models, *B* and *C*, develop, both of which describe the following subsystems of the earth's climate: atmosphere dynamics, ocean dynamics, ice formation, and the effects of vegetation. Model *B* develops from *A* via the following path: first, a module for ocean dynamics is coupled to Model *A*, followed by a module for ice formation, and finally a module for the effects of vegetation. Model *C* takes a slightly different path; first, a module for ocean dynamics is coupled to *A*, followed by a module for the effects of vegetation, and finally a module for ice formation. Because B and C took different developmental paths, the attempt to couple additional modules to B and C—for example, a module for aerosols—will, in general, need to proceed along different lines for each model: the piecemeal approximations to parameterization schemes that will allow a new module to be coupled to B will, in general, not work for C, and vice versa. This does not mean that it will be impossible to add an aerosol module to C; however, given the ways in which B and C have developed, it might be significantly more difficult to add this module to C than to B, and it almost certainly will need to be done via a very different coupling process. These difficulties, moreover, only increase with the complexity of the model in question. Thus, given models of the complexity that we have today, the historical development of these models—which, again, has been influenced significantly by the "environmental" pressures placed upon them—places constraints upon the kinds of modules that can be coupled to them, which in turn affects the overall performance of the models.

The preceding discussion should make clear why attribution has been, and will in all likelihood continue to be, unsuccessful. One cannot attribute the predictive success or failure of a given model to a particular, localized component of that model because the components of models are strongly coupled to one another and hence interact with one another in significant fashion. Moreover, the kinds of additions that one can make to a climate model—for example, the ways in which we can expand the scope of a model—will in general depend upon the way in which that model has developed over time, including the approximations and adjustments that have been made in order to couple new modules successfully. This, in turn, will depend upon the "environmental" pressures to which the model is subject. All of this suggests that previous decisions to emphasize certain prediction and retrodiction tasks over others—for example, predictions of global mean surface temperature change over predictions of global precipitation change—have a significant effect upon the ways in which these models can be expanded, and upon the ways that future expansions of these models will perform.

Values and the Estimation of Structural Model Uncertainty

What does all of this have to do with the debate over the role of values in scientific research? According to the preceding discussion, the overall performance of a model—its ability to predict and retrodict certain tasks well, and its inability to predict and retrodict other tasks well—depends to a significant extent upon the history of that model, which in turn depends upon that model's "environment," which includes political decisions to emphasize certain predictive and retrodictive tasks over others. Because the estimation of uncertainties in our knowledge of climate change depends upon the performance of our best climate models, and because decisions of a non-epistemic character have an ineliminable effect upon the performance of our best climate models, such non-epistemic values invariably influence the estimation of uncertainties. Thus, Jeffrey's claim that we can assign probabilities to hypotheses in a value-neutral fashion—or, in the present case, that we can assign uncertainties to predictions in a value-neutral fashion—is false, at least in the area of climate modeling.

One might object to this argument in a few different ways. First of all, our argument depends upon the claim that the choice of a prediction task is inevitably influenced by non-epistemic factors; yet one might argue that the decision to emphasize temperature predictions over other predictions can be explained solely on theoretical, or "purely epistemic," grounds. While one could develop this objection in a variety of ways, perhaps the most plausible would proceed as follows. Of the candidate prediction tasks that one could choose to emphasize—including global mean surface temperature change, ocean heat content change, polar ice cap retreat, and so on—temperature is the most significant theoretically, because all of the other quantities can be derived from it. Without knowing how global mean surface temperature will change, so the argument goes, one cannot explain or make sense of any other change in global climate; therefore, predictions of temperature change should be emphasized.

The objection thus formulated, however, has its problems. It is true that CO_2 and other gases are often called "heat trapping," but it would be more accurate to call them "energy trapping," and one of the most difficult problems within contemporary climate research is determining where this energy will go. Will it go toward heating the ocean? melting sea ice? or raising temperature? If it goes toward the latter, will it do this on the surface of the earth or in other parts of the atmosphere? To maintain that all of these other features of the planet's dynamics can be *derived* from mean global surface temperature is misguided, because it fails to take into account the massive and not entirely understood interdependences that exist in the climate. All of these interdependencies together react in complex fashion to the trapping of additional energy, and it would be a mistake to assume that this additional energy will always manifest itself as higher surface temperatures. Increased surface temperature is only one of many symptoms of global climate change.

This seems to be the most plausible way of developing this objection, and yet it still fails. One could, of course, attempt to formulate it in another fashion. I am skeptical, however, that such a route will be successful. Until this burden is met, therefore, I conclude that the decision to emphasize predictions of global mean surface temperature change is ineliminably influenced by non-epistemic considerations, such as the fact that this prediction task was seen as the most important in terms of its social, political, and economic consequences.

A second objection that one might direct against our argument is that the way in which values operate in this case is philosophically uninteresting; if scientists decide, for whatever reason, to put more resources into the prediction of one task than another, it is hardly novel to maintain that the resulting models will be better at predicting the former task than the latter. This objection, however, fails for at least two reasons.[17]

First of all, the objection obtains its intuitive force via a false presupposition, namely, that it is realistic to attempt to construct models that perform in a manner that is independent of the influence of value judgments. According to the previous discussion, the performance of current climate models depends *invariably* upon the historical development of those models, which in turn depends *invariably* upon decisions of a social or political character, including decisions that a given prediction and retrodiction task is more important, from a social, moral, or economic point of view, than another. If it were reasonable to expect the performance of climate models to be independent of such past judgments, then the objection would be a good one; according to our argument, however, this expectation is not at all reasonable.

Second, and perhaps more important, the objection misrepresents the effects of focusing upon temperature predictions at the expense of precipitation predictions. In particular, the objection suggests that the primary effect of focusing upon one kind of prediction task rather than another is a difference in the actual uncertainties associated with these two predictive tasks. Yet, it is not merely the case that a lack of focus upon precipitation predictions will result in models that perform poorly with respect to this prediction task—and thus lead to higher actual uncertainties about future precipitation; it is also the case that a focus on mean global surface temperature predictions at the expense of global precipitation predictions will likely result in a range of models that *overstates* our estimates of the uncertainty of precipitation predictions and *understates*

---

17. Thanks to James McAllister for encouraging us to think further about this objection.

the uncertainty of temperature predictions. This choice of prediction tasks, in other words, affects not only the actual uncertainties of the models in question, but also the *estimations* of these uncertainties.

The reason for this pertains to the problems associated with estimating structural model uncertainty via observable frequencies—or, more specifically, via observing the range of predictions generated by our existing arsenal of models. If there are some particularly bad models on the market, we will overestimate the degree of uncertainty that we ought to have about our best models. On the other hand, if there is some degree of coevolution in our models—that is, if those who make them deliberately ensure that the predictions of their models do not deviate too much from the herd (a natural thing to do if you do not want the predictions of your model to appear unrealistic)—then the range of models on the market will underestimate the degree of uncertainty that we ought to attribute to them. As Myles Allen, who refers to this method as using "ensembles of opportunity," notes, "If modelling groups, either consciously or by 'natural selection,' are tuning their flagship models to fit the same observations, spread of predictions becomes meaningless: eventually they will all converge to a delta-function" (Allen 2008). Thus, by misrepresenting the effects of focusing upon temperature predictions at the expense of precipitation predictions, the objection obscures the significant impact that values have in this situation.

Structural Model Uncertainty and Expert Judgment

Of the four modalities that we discussed earlier (expert judgment versus observable frequencies and structural model uncertainty versus parameter uncertainty), we have thus far restricted our attention to the estimation of structural model uncertainty via observable frequencies. The aforementioned problems with this method, however, lead to a further challenge to our claim that non-epistemic considerations play an ineliminable role in the estimation of structural model uncertainty—namely, that one could perhaps make such estimations via expert judgment rather than observable frequencies and do so in a manner that is free from non-epistemic values. Are there reasons for believing that values play an ineliminable role in the case of expert judgment as well?

I believe that there are such reasons, but the question is nevertheless rather difficult to answer, primarily because, of the four modalities under consideration, the use of expert judgment to estimate structural model uncertainty is the one with the least developed methodology in the literature.[18] The source of this lacuna is clear; when statisticians solicit the opinions of experts regarding the degree of structural uncertainty in a model, they often encounter the problem that the experts are either unwilling or unable to assign subjective degrees of belief to the structure of a model. In particular, modelers are extremely reluctant to assign probabilities to the predictions of models that are deterministic, given a particular value in parameter space.[19] When statisticians, trying to implement this modality, make it clear that what they are after is the modelers' subjective degree of belief, the modelers are apt to balk. Thus it is not clear that the relevant experts actually possess the subjective degrees of belief, and methods need to be created that will reliably elicit these judgments.

The issue that is of interest to us, of course, is the extent to which experts' judgments about their subjective degrees of belief, assuming that they are willing to make these judgments, are influenced by past decisions about prediction tasks. Given, again, that a well-established methodology for making these judgments does not yet exist, the issue is difficult to resolve. We can, however, make the following preliminary argument. In order for expert judgment to overcome the effects that past prediction-task priorities have had on ensembles of existing models, the experts in question must understand exactly what these effects have been. Yet, the failure of the project of attribution, discussed above, leads us to be skeptical that one can know what the precise effects of past prioritizations upon existing models actually are. This, in turn, makes us doubt that expert judgment can be free from non-epistemic considerations, such as the choice of one prediction task over another.

Where does this leave us? We have argued that estimations of structural model uncertainty via observable frequencies cannot be made in a value-free way, and we have called into question the claim that expert judgment can do so. In order to establish this latter point more firmly, more research on the methodology of this modality needs to be done. In the meantime, we have to be content with the fact that the predominant method in actual use in the scientific community is to look at observable frequencies; this method, as we have argued, is not value-free.

Suppose, however, that one is unconvinced by the arguments just provided. Even if one continues to maintain that structural model uncer-

18. For a discussion of these methods, see Goldstein and Rougier 2006. For criticism of these methods, see Allen 2008.

19. Modelers, in fact, often think statisticians will do that for them. Statisticians, on the other hand, clearly (and correctly) view this as a scientific problem whose answer can only be provided by those with the best scientific expertise.

tainty can be estimated independently of non-epistemic considerations—indeed, even if one assumes that there is no uncertainty whatsoever about the structure of a model—there is still significant uncertainty about the values of parameters and parameterization schemes. I believe that an even stronger case can be made for the claim that non-epistemic considerations affect the estimation of parameter uncertainty. It is to this argument that we now turn.

Parameter Uncertainty

To keep the issue of parameter uncertainty conceptually distinct from that of structural model uncertainty, let us suppose that we do have a structurally perfect model with n parameters and that we are uncertain about what the best value is for each of these parameters. We can think of the set of n parameters as forming an n-dimensional parameter space. If we had a probability density function (PDF) over that n-dimensional space, we could easily assign probabilities to the predictions of the model. In principle, we would use something like the following Monte Carlo method: we could sample from the space of parameters in accordance with the given PDF and calculate the output of the model for each of the sampled points in the space.[20] The resulting set of outputs would have means and variances for each of the variables that would correspond to the uncertainties we would assign to the corresponding predictions. If, for example, the mean prediction of the probability-weighted sample for temperature is x, with standard deviation of s, then we could say that it is 95 percent likely that the modeled system's temperature will be $x \pm 2s$.

The question, then, is where a PDF over the space of possible parameter values can come from. The answer, as noted, is from observed frequencies and expert judgment. We have already seen, in some detail, how the former works. Each value in parameter space can be benchmarked; it can be scored for its ability, when used as input for a given model structure, to reproduce existing data, and probability densities can be assigned in proportion to that score. Remember, however, that climate models are highly multivariate in their output. Scoring the performance of each point in parameter space, therefore, requires us to weight the relative

---

20. I say "in principle" because in practice the situation is more complicated. In practice, it is too time-consuming to calculate the outputs of a global climate model for sufficiently many sampled points. Statisticians, therefore, use so called "emulators" of the models (in effect, simulations of the simulations), to calculate these outputs. While this introduces a large layer of technical difficulties to the project, it has no bearing, that we are aware of, on what we discuss here.

importance of making good predictions of each variable. Such a weighting involves a value decision that involves both epistemic and non-epistemic considerations.[21] More important, expert judgment might affirm that a point in parameter space that is very successful at reproducing known, available data will be poor at predicting unknown, unavailable data. In practice, therefore, a more common procedure is to use a benchmarking procedure to rule out certain regions of parameter space (if a point in parameter space does a very poor job of reproducing known data, then it can be thrown out) and then use expert judgment to assign a PDF to the remaining subset of the space.

But what are experts judging when they assign such a PDF? Here, we need to keep in mind an important difference between two kinds of parameters that climate models can take as inputs. So far, I have simply defined a parameter of a model as any aspect of the model that has to be quantified before the model simulator can be run. But the aspect of the model being quantified might or might not correspond to some actual quantifiable or measurable property of the physical system being modeled. While some of the parameters of our models are of a kind with physical parameters like g, the acceleration due to gravity near the earth's surface (9.8 m/s^2), others of them, which result from so-called parameterizations in the model, are rather different.

Parameterizations are elements of a simulation model that are designed to capture effects that slip between the cracks of a model's discretization grid or are otherwise lost to an approximation of the model. Parameterization schemes are extremely common in global climate models, especially since reductions in discretizations and approximations are bought at the price of complexity of computer models, increase of simulation time, and so on.[22]

In my view, there is an important difference between these two different kinds of parameters. With respect to the first kind of parameter, it makes sense to talk about the correct value of the parameter. For example, near the surface of the earth, the correct value for g is approximately 9.8 m/s^2. But the value of a parameter associated with a parameterization scheme does not have a single correct value. At best, it has a best value

21. Here, we think, is an interesting example of where the purported distinction between epistemic values and non-epistemic values breaks down. Whether or not the value underlying such a weighting would count as epistemic or non-epistemic would depend on whether the researchers could argue that they made their choice of weighting on the basis of which weighting was the "best guide to truth." But which truth? And how would such an argument be resolved? The distinction here becomes a bit confused.

22. See chapter 1 for more discussion of parameterizations.

for a particular prediction task—a value which, if used as model input, will enable the model to make the best possible predictions for that particular task.

Goldstein and Rougier highlight the important difference between these two kinds of parameters when it comes to estimating uncertainties. In their discussion of eliciting expert judgment about appropriate PDFs to place on parameter space, they write, "In practice, modellers often seem to take two somewhat contradictory positions about the status of the simulator's best input, on the one hand arguing that it is a hypothetical construct and on the other hand using knowledge and intuition derived from the physical system to set plausible intervals within which such a value should lie" (Goldstein and Rougier 2009, 1226).

Of course, when it comes to parameterization schemes, there is no straightforward relation between the values of the inputs to the simulator and any measurable corresponding physical values for the system. If expert judgment is to be relied upon in assigning probabilities to parameter values, therefore, it has to come not from *knowledge and intuition of the physical system alone*, but from knowledge and intuition of the behavior of the model vis-à-vis the system. In that regard, one more passage from the Goldstein and Rougier piece is especially striking: "In particular, there was strong disagreement [among statistically numerate system experts] with [the following:] that there exists an input x_* such that, were it to be known, only a single evaluation of the simulator would be necessary" to learn everything that the model has to tell us about the system (2009, 1226). What the statistically numerate system experts are suggesting is that the best value from parameter space for one particular prediction task is not necessarily the same as the one for another prediction task. Indeed, it may even be that for some prediction task, the maximum amount of knowledge that can be extracted from a particular model structure might come from an ensemble of runs from multiple parameter values. Thus, the idea that there is a single "best value" for a parameter in a model is, in many cases, not correct.

What is important to note here is that, as emphasized in the previous section, climate models can be applied to a variety of prediction tasks. While climate modelers have traditionally emphasized predictions of global mean surface temperature change, they have more recently begun to turn their attention to other prediction tasks, such as the probability of so-called abrupt climate changes, including the collapse of the thermohaline circulation, the drying of the Amazon, and the disappearance of Arctic summer sea ice. What we do not know for a particular climate model with all of its parameterizations, is whether the best parameter value for predicting mean surface temperature is also the best parameter value for predicting, say, the extent of thermohaline circulation. But if the claim made by Rougier and Goldstein's statistically numerate system experts is right (and I think it is), and they are not always the same, then the following suspicion arises.

Suppose we are confronted by a novel prediction task, such as estimating the probability of thermohaline collapse. And suppose that, for a particular model structure, we want to estimate the degree of parameter uncertainty about such predictions. To do this, as we have discussed, would require us to use expert judgment to assign a PDF to the parameter space for that model. If what experts are judging is where, in parameter space, the "best" value of parameters can be found, and if the best value for predicting mean surface temperature is not necessarily the best value for predicting thermohaline circulation, then the suspicion is that expert judgment about parameter space PDFs will be influenced, to some degree or other, by the intended purposes for which these experts have been constructing and testing their models during the period in which they have acquired their expertise. Again, when it comes to judgment about parameterizations, the relevant expertise is not about the system itself, but about the behavior of the model vis-à-vis the system.

Values and the Estimation of Parameter Uncertainty

At this point, we are in a position to state the argument for the ineliminability of non-epistemic considerations in the estimation of parameter uncertainty. Suppose that we have a particular model structure about which there is no uncertainty, and suppose furthermore that we want to employ this model in order to estimate the probability of thermohaline collapse given a doubling of atmospheric carbon dioxide. To estimate this, we have to elicit expert opinion regarding an appropriate PDF to assign to the space of parameters. But the best value(s) of parameters for predicting thermohaline circulation is not necessarily the same as the best value(s) for other prediction tasks, such as predicting mean surface temperature. If the experts whose opinion we elicit have predominantly acquired experience with their models in predicting mean surface temperature, then this will affect their judgment about where the best value for parameters can be found.

But the particular set of prediction tasks that have played a role in shaping our experts' judgments have been the product of a set of choices—for example, the choice to focus on predicting mean surface temperature rather than mean global precipitation. And these choices, in turn, reflect a set of values—namely, the set of social, economic, or other considerations that have historically led us to believe that predicting temperature is more important than predicting precipitation. These values, in other words, are ones that are traditionally regarded as being non-epistemic in character. Thus, non-epistemic values play an ineliminable role in the estimation of parameter uncertainty.

Conclusion

Jeffrey argues that the task of the scientist can and should be limited to the assignment of probabilities to hypotheses, and that this task can be carried out in a manner that is free from practical, non-epistemic considerations. Once the task of assigning probabilities to hypotheses has been completed, the scientist can then hand these hypotheses and probabilities over to policymakers, who are responsible for deciding upon a course of action. On this picture, there is a clean separation between the realms of theory and practice, and a clear line that divides the spaces where values play a legitimate role (the realm of practice) and where they do not (the realm of theory). As noted earlier, Jeffrey's line of reasoning is still very commonly followed today.

If our argument is sound, however, Jeffrey's line of reasoning fails, at least in one very important area of contemporary scientific research—namely, climate modeling. Scientists cannot assign probabilities to hypotheses about climate change—or, more specifically, estimate the uncertainties of climate predictions—in a manner that is free from non-epistemic considerations, because non-epistemic considerations invariably influence the choices of prediction tasks, and the choices of prediction tasks invariably influence the estimation of both structural model uncertainty and parameter uncertainty.

Again, I do not believe that this result in any way implies that the consensus that has been formed regarding the causal connection between fossil-fuel emissions and global climate change is problematic. I am not a skeptic regarding the fundamental claim of anthropogenic climate change. I do, however, believe that this conclusion suggests that more attention should be paid to the spaces within climate modeling where values play a role, to the kinds of values or non-epistemic considerations that play a role, and to the effects that these values have upon the overall performance of our models.

SEVEN

Reliability without Truth

One of the principal lessons of this book has been that the mathematical models that drive many computer simulations of complex physical systems have complex ancestries. On the one hand, the basic features of these models are motivated by fundamental theory. On the other hand, in order to produce a model that is computationally tractable, simulationists also craft their models using a motley assortment of other components, incorporating many assumptions that are not sanctioned by high theory. Despite their mixed ancestries, many of these simulations are trusted in making predictions and building representations of phenomena, and they are often successfully used in engineering applications. Indeed, researchers run simulations of systems about which data from real experiments is difficult or impossible to get—the simulations take the place of experiments and observations—and so they are trusted even in circumstances where they cannot be evaluated by comparing their results with the world.

Given that the construction of these models is guided but not determined by theory, what is the source of the credibility of these models? In preceding chapters, I have argued that the credibility of a simulation model must come not only from the credentials supplied to it by its theoretical ancestors, but also from the antecedently established credentials of the model-building techniques employed in its construction. There are, in other words, model-building techniques that are taken, in and of themselves, to be suitable for building credible models. Some of these techniques, moreover, go beyond idealization or approximation and involve incorporating what can be best described as fictions. These are submodels that are included in a simulation model; their inclusion is taken to increase the trustworthiness of the simulation's output, but their function is not thought to be to offer a reliable account of the bits of the world they purport to model.

The practice of using fictions in building credible simulations is worthy of closer scrutiny by philosophers of science interested in the various arguments for and against scientific realism. Here, I examine two examples of fictions from the field of computational fluid dynamics: so-called artificial viscosity and vorticity confinement. Both of these techniques are successfully and reliably used across a wide domain of fluid dynamical applications, but both make use of "physical principles" that do not purport to offer even approximately realistic or true accounts of the nature of fluids. I argue that these kinds of model-building techniques, therefore, are counterexamples to the doctrine that success implies truth—a doctrine at the foundation of scientific realism. I suggest, furthermore, that fictions can provide a useful backdrop for thinking more carefully about what characteristics we take successful model-building principles to have.

Autonomy

We should begin by recalling two important features of simulation research. The first is what I call, following Mary Morgan and Margaret Morrison (1999), their semi-autonomy from theory. While models generally incorporate a great deal of the theory or theories with which they are connected, they are usually fashioned by appeal to, by inspiration from, and with the use of material from an astonishingly large range of sources: empirical data, mechanical models, calculational techniques (from the exact to the outrageously inexact), metaphor, and intuition.

A second important feature of simulations is that they are often constructed precisely because data about the systems they are designed to study are sparse. In these circumstances, simulations are meant to replace experiments and observations as sources of data about the world. Simulation methods, for example, are used to study the inner convective structure of stars or to determine the distribution of pressure and wind speed inside a super-cell storm. Not all of the results of such simulations can be evaluated simply by being compared to the world. If a simulation reveals a particular pattern of convective flow inside a star, we must be able to assess the trustworthiness of that information without being able to physically probe the inside of the star to check and see whether that result is confirmed by observation. In this sense, we can speak of simulations as being independently sanctioned. By this, I simply mean that if a simulation is to be useful, it must carry with itself some grounds for believing in the results it produces. The process of transformation itself, from theoretically given model to computationally tractable model, must be sanctioned. This is a feature of simulation that I have previously highlighted by talking of its "downward epistemology" (Winsberg 1999).

When these two features of simulations—their semi-autonomy from theory and the fact that they are independently sanctioned—are held up side by side, it raises an interesting question. If the process of constructing simulation models is at best only guided by theory, then how can the simulation be trusted to produce results in situations where data are sparse? What, other than the governing theory, could provide the necessary credentials?

Part of the answer, which we saw in chapter 3, is that the techniques that simulationists use to construct their models are "self-vindicating" in much the same way that Ian Hacking says of instruments and experiments that they are self-vindicating (Winsberg 2003). That is, when simulationists build a model, the credibility of that model comes not only from the credentials supplied to it by the governing theory, but also from the antecedently established credentials of the model-building techniques used to make it.

The principle purpose of simulations is to produce "results." These results come in the form of simulated "data" that are expected to, in specifiable respects, accurately predict or represent the phenomena to be simulated. Whenever these techniques and assumptions are employed successfully—that is, whenever they produce results that fit well into the web of our previously accepted data, our observations, the results of our paper-and-pencil analyses, and our physical intuitions; whenever they make successful predictions or produce engineering accomplishments—their credibility as reliable techniques or reasonable assumptions grows. That is what I meant when I said that these techniques have their own life; they carry with them their own history of prior successes and accomplishments, and, when properly used, they can bring to the table independent warrant for belief in the models they are used to build. In this respect, simulation techniques are much like experiments and instruments as Hacking and Peter Galison describe them (Hacking 1988; Galison 1997).

That was the answer I gave to the question, "What makes it possible for semi-autonomous simulation models to be credible sources of knowledge about systems for which data are sparse?" That is an epistemological question. Here I want to explore some of the implications of the answer to that question for the metaphysics of science. Roughly, I want to ask if there are any lessons we can learn from the observation that model builders develop trust in the success of some of their techniques, even when their techniques employ contrary-to-fact physical principles.

Asymmetries of Success

Let me begin to motivate the idea that this observation might have some interesting implications. Recall the following claim. We sometimes trust the results of simulation models because

- we place trust in the theories that stand behind these models. We do this, of course, because they have been successful in lots of other applications,
- and, equally, because we place trust in the model-building techniques that we use to transform theoretical models into simulation models. Here too, we do this because those techniques too have been successful in lots of other applications.

This way of putting things makes it seem as though there is a perfectly simple symmetry at work here, but the apparent symmetry obscures a significant difference. We all know that theories and laws are the sorts of the things that are supposed to gain credibility—to be corroborated— every time they are applied successfully. But I take it that it might come as more of a surprise to some that a model-building technique would be the right sort of thing in which to develop confidence. Although a model-building technique, a particular way of altering a theoretical model so as to make it more computationally tractable, may produce good results in one particular application, we might be tempted to ask why we should expect it to work in another application. Why, in other words, should the success of a model-building technique be the sort of thing that is projectable from one application to another?

The reason I think this is an interesting question is that it is often thought that there is only one possible explanation for our confidence in the projectability of scientific success. If we ask why we expect scientific theories that have been successful in past applications to be successful in the future, we often get a response that goes something like this: If a proposed theory or law is used successfully in making a variety of predictions and interventions, then it is likely that that theory or law is in some way latching on to the real structure of the world—that it is true—and hence it is only natural that we should expect it to be successful in the future. If we did not believe it was latching on to the real structure of the world, so the reasoning goes, then we would have no reason to expect it to be successful in the future. In short, the success of scientific theories is projectable because successful scientific theories are true.

In fact, many think that the truth of scientific theories is the *only possible explanation* of their success in making predictions and interventions. Otherwise, they claim, that success would be a miracle. Many also think that our *belief* in the truth of scientific theories is the only possible *explanation* for our common practice of trusting theories in making predictions about novel situations. These, of course, are standard arguments for scientific realism. But what, then, explains our practice of trusting *model-building techniques* in doing the very same thing?

Arguments for scientific realism, in other words, rest at least in part on the *conviction* that the projectability of the success of scientific theories calls for an explanation, and that the only possible or viable explanation available is truth. In what follows, I ask whether, in light of the fact that some aspects of scientific practice *also* seem to rely on the projectability of the success of model-building techniques, this conviction is warranted.

No Miracles

The idea that the success of scientific theories requires an explanation and that the best explanation is truth forms the basis for what Philip Kitcher has called the "success-to-truth" rule (Kitcher 2002), which in turn is the engine of the "no-miracles" argument for scientific realism. I state the rule crudely below. In the next section, following Kitcher, I review some considerations that force upon us a more nuanced formulation.

If *X* plays a role in making successful predictions and interventions, then *X* is true.

Of course, the no-miracles argument and the rule of inference on which it depends have many critics. One kind of criticism questions whether correspondence-truth really has any explanatory value to begin with (e.g., Horwich 1999). That avenue of criticism will not concern us here. Other critics have looked for counterexamples to the principle (e.g., Laudan 1981), usually in the form of historical examples of scientific theories that were successful but that we no longer hold to be true—such as the humoral theory of disease or the wave-in-ether theory of light.

Defenders of the no-miracles argument have responses to such counterexamples that invoke qualifications to the rule. Take, for example, the case of the humoral theory of disease. Examples of this type are often described by defenders of the rule as belonging to "immature science." The response, in other words, is that in order for X to be a genuine example of something to which the rule ought to apply, it needs to play a role in making *sufficiently specific and fine-grained* predictions and interventions. Because, the argument goes, such theories as the humoral theory of disease and the phlogiston theory of combustion never allowed for sufficiently specific and fine-grained predictions and interventions, they are not good counterexamples to a properly formulated rule. Our first modification to the rule, therefore, must accommodate this concern.

Another canonical counterexample to the no-miracles rule is the wave-in-ether theory of light. So-called structural realists have taken this example to be paradigmatic of a certain kind of counterexample and have crafted their version of realism in response. What they have urged is that these historical examples of successful but untrue theories can be divided into two parts: (a) a part that is no longer taken to be true but that did not play a genuinely central role in the relevantly successful predictions and interventions, and (b) a part that did play a genuinely central role, but that is still taken to be true. In the case of the ether theory, (a) is the ontology of the theory (namely, the existence of the ether), which they argue did not play a genuinely central role in making predictions or interventions, and (b) is the mathematical form of the theory, which they argue can still be regarded as true. So while structural realists admit that many scientific theories have been successful without being true in their entirety, they argue that the successes of these theories have been due entirely to the fact that one component of the theory, its structure, has accurately reflected reality. Structural realists, in other words, argue that the rule of inference should be modified so that it only applies if X, *in its entirety*, plays *a genuinely central role* in making the successful predictions and interventions.

It has also been widely recognized that many successful theories and models, both from the past and present, cannot be held to be literally true or to represent exactly. The model of a simple harmonic oscillator can quite successfully predict the behavior of many real physical systems, but it provides at best only an approximately accurate representation of those systems. Newtonian mechanics is a very successful theory, but it is at best a limiting case of a true theory. Even maps can be successfully used to navigate their intended territories, but they omit many details and distort certain features. Because of these considerations, many defenders of the no-miracles rule accept that it needs to be modified so that it guarantees not the truth of *X tout court* but truth in some qualified sense—for example, the approximate truth of *X*, or the fact that *X* is a limiting case of something true. For now, we can rewrite the consequent of the rule to read: "*X* is (*in some qualified sense*) true."

Ad hoc hypotheses are another class of obvious exceptions to the no-miracles rule. In order to qualify for application in the rule, it is widely recognized that *X* must achieve success across a wide range of applications. It is even better if the success of *X* is projectible—that is, that we not only think *X* can be used for some domain for which it was designed, but that we have the expectation that it will be useful in future domains. In short, the success of *X* has to be *systematic* (as opposed to ad hoc).

One final modification to the rule as stated above is required. It is fairly obvious that the scope of the variable *X* in the rule cannot range over all entities. No one would deny that a calculator, a triple-beam balance, and even a high-energy particle accelerator can all play genuinely central roles in making specific and fine-grained predictions and interventions. But no one would want to have to defend the view that any of these entities is "true," even in any qualified sense. So, to be pedantic about it, we need to be perfectly clear that the no-miracles rule applies only if the *X* in question is the right sort of entity to be a candidate for truth and falsity, or, at the very least (in order to include such things as maps and models), to exhibit similarity of structure with the world. In short, *X* should be a representational entity of some sort.

A properly formulated success-to-truth rule of inference thus reads as follows:

If . . .

(the right sort of) *X* (in its entirety) plays a (genuinely central) role in making (systematic) successful (specific and fine-grained) predictions and interventions.

Then . . .

X is (with some qualification) true.

Thus, if some *X* is going to be offered as a counterexample to the no-miracles rule in a way that advances the debate about that rule, then it had better be the case that

- *X* plays a genuinely central role in making predictions and interventions.
- Those predictions and interventions are specific and fine-grained.

- *X* cannot be separated into a part that is false and a part that does the relevant work.
- The predictions and interventions we use *X* to make occur across a wide range of domains, and we sometimes confidently apply *X* in new domains.
- *X* is the relevant sort of representational entity.
- *X* cannot plausibly be described as true, even in some suitably qualified sense.

I want to argue here that the field of computational fluid dynamics offers some plausible candidates for counterexamples to the no-miracles rule that meet all of the above criteria. In what follows, I discuss two of them: "artificial viscosity" and "vorticity confinement."

The Success of Fictions

One of the earliest uses of finite-difference simulations arose in connection with the Manhattan Project during World War II. John von Neumann and his group of researchers used finite-difference computations to study the propagation and interaction of shock waves in a fluid, a subject crucial to the success of the atomic bomb.

We generally think of shock waves as abrupt discontinuities in one of the variables describing the fluid, but it was quickly recognized that treating them in this way would cause problems for any numerical solution. The reason is that a shock wave is not a true physical discontinuity, but a very narrow transition zone whose thickness is on the order of a few molecular mean-free paths. Even with today's high-speed and high-memory computers, calculating fluid flow with a differencing scheme that is fine enough to resolve this narrow transition zone is wildly impractical. However, it is well known that a simulation of supersonic fluid flow that does not deal with this problem will develop unphysical and unstable oscillations in the flow around the shocks. These oscillations occur because of the inability of the basic computational method to deal with the discontinuities associated with a shock wave—the higher the shock speed, the greater the amplitude of these oscillations becomes. At very high speeds, such a simulation quickly becomes useless. To make it more useful and accurate, simulationists somehow need to dampen out these oscillations.

The generally accepted way to do this, which was originally devised by von Neumann and Richtmyer while working at Los Alamos (von Neumann and Richtmyer 1950), is to introduce a new term, an "unphysically large value of viscosity" into the simulation, which is called *artificial viscosity*.[1] The inclusion of this term in the simulation is designed to widen the shock front and blur the discontinuity over a thickness of two or three grid zones. The trick is to apply this viscosity only to those portions of the fluid that are close to the shock front. This is achieved by assigning a magnitude to the fake viscosity that is a function of the square of the divergence of the velocity field—which happens to be a vanishing quantity everywhere but close to the shocks. The end result is that the method enables the computational model to calculate certain crucial effects that would otherwise be lost inside one grid cell—in particular, the dissipation of kinetic energy into heat.

Artificial viscosity is not the only nonphysical "effect" used in simulations of physical systems. Another example from computational fluid dynamics is what is known as "vorticity confinement" (Steinhoff and Underhill 1994). The problem to be overcome in this case arises because fluid flows often contain a significant amount of rotational and turbulent structure at a variety of scales. When that structure manifests itself at scales too small to be resolved on a grid of the size used in the simulation, significant flow features can become "damped out." This undesirable effect of the differencing scheme is called "numerical dissipation." The solution is to use a technique called vorticity confinement.[2] The method consists in finding the locations where significant vorticity has been numerically damped out and adding it back in using an artificial "paddle wheel" force. Much as in the case of artificial viscosity, this is all done by a function that maps values from the flow field onto values for the artificial force.

For the rest of the discussion, I confine my remarks to the example of artificial viscosity. Most of what I say, however, can be repeated mutatis mutandis, about the paddle wheel force, as well as, presumably, about a variety of other fictions used in simulation, including, of course, silogen atoms. Artificial viscosity is simply the oldest and most established of these techniques. Some remarks about artificial viscosity are now in order.

- Artificial viscosity is clearly a successful tool of scientific investigation, prediction, and intervention.

---

1. For a modern discussion of artificial viscosity and its applications, see Campbell 2000 and Caramana, Shashkov, and Whalen 1998. The original presentation appears in von Neumann and Richtmyer 1950.
2. See Steinhoff and Underhill 1994.

- The success of artificial viscosity, furthermore, is systematic and projectable. It is "projectable" in the sense that physicists and engineers use this technique of computational fluid dynamics to make *novel predictions* about flows containing shocks—its inclusion in a simulation of flows with strong shocks adds to the confidence that researchers have in their results. The success is "systematic" in the sense that its use has been studied extensively, and there is a wide body of knowledge about how to use this modeling principle effectively in an off-the-shelf manner. It is used, furthermore, in an enormous variety of applications ranging from engineering applications to astrophysics.
- Artificial viscosity is not like the humoral theory of disease. It is used to make very fine-grained and detailed predictions, descriptions, and interventions. Without it, the Manhattan Project might not have succeeded.
- Artificial viscosity is not like the claims about the ether. It can perhaps be said of the nineteenth-century wave theory of light that the claims it contained about the ether were superfluous—that these claims "could be dropped from the theory without affecting the success of the practice." This is clearly not the case with artificial viscosity. Artificial viscosity plays a crucial role in damping oscillation instabilities that would otherwise render simulation results useless.
- I said earlier of simulation *techniques* that they can bring to the table independent warrant for belief in the models they are used to build. There are, in short, such things as widely successful model-building techniques. This fact alone, however, might not worry the proponent of the no-miracles rule, because *techniques* are not the sort of things that are candidates for being true or false, or for providing accurate representations. But artificial viscosity does take the form of a claim. We can easily think of artificial viscosity as the claim made of fluids under its domain that they display a viscosity that is proportional to the square of the divergence of their velocity field. In this sense, artificial viscosity is indeed a candidate for truth or falsity. The same is true of the technique of vorticity confinement. The technique calls for the application of the physical principle that certain kinds of fluid flows give rise to a paddle wheel force that arises in proportion to certain characteristics of the flow. This principle is, as well, a candidate for truth or falsity.

Finally, we need to address the question of whether we can understand X to be (perhaps in a qualified sense) true. To begin to do this, we need first to distinguish the model that drives the simulation, the principles that go into it, and the results that the simulation produces. Recall that we consider a simulation model to be successful if we have reason to think that predictions and representations of the phenomena that it produces—its results—are accurate in the respects that we expect them to be. Nothing whatsoever about my argument should prevent us from understanding these simulation results in realistic terms if we so choose.

Things are slightly more complicated when we look at the models that drive the simulations. Since these models incorporate false assumptions, we cannot view in them in literally realistic terms; they surely do not offer literally true accounts of the actual functional relationships that exist between the various properties of the fluid. But we are concerned here not only with literal truth, but with the possibility of approximate truth.

I take it, after all, that this is for many philosophers the standard way to think about successful contemporary theoretical structures that, for whatever reasons, do not seem like plausible candidates for being true descriptions of the world. The fact that they are successful is taken to be evidence that they must approximate some (perhaps as-yet-undiscovered) ideal theory.

Approximate truth is a slippery subject. There is nothing like widespread agreement among philosophers about what a theory of approximate truth should look like, or even, for that matter, if such a theory is even desirable or possible. Luckily, we can set these worries aside for our purposes and ask only whether it is even plausible to think that the principle of artificial viscosity would come out as approximately true on any account of approximate truth. A separate difficulty, on the other hand, arises from the fact that there are two different ways to think about the question in this case: one that focuses on the models that drive the simulations, and the other on the principles that inform the model construction.

One way to ask the question is this: Should we be willing to say that there might be *some qualified sense* in which students of fluid dynamics accept as true (or should accept as true) the claim that certain fluids display a viscosity that is proportional to the square of the divergence of their velocity field? Here, I think the answer is clearly no.

Viscosity, in fluid dynamics, is the measure of how resistive the fluid is to flow. Most fluids can be well modeled by assuming that the sheer stress between parcels of fluid is proportional to their relative velocities and some constant, called the viscosity, which is taken to be a physical property of the fluid itself. These are called "Newtonian" fluids. In very high Reynolds number flows, like the kinds of flows we have been talking about, viscosity often becomes an insignificant parameter—it is often left out of the model. Of course, no one really thinks that any fluid is truly Newtonian, and certainly no fluid is inviscid. But treating a fluid as Newtonian, or inviscid, might arguably be thought of as providing a good approximation to the actual forces that small parcels of fluid experience as they slide against each other. On the other hand, when simulationists refer to "artificial viscosity" and label it an "unphysically large, viscosity-like term," they are signaling that the term in the equation is not meant to capture, not even approximately, relationships between properties of the fluid and the forces that occur within them.

Defenders of the idea that success implies truth, however, might want to argue that this is the wrong question to ask. The relevant sort of question, one might argue, is not whether the claim that certain fluids display a viscosity that is proportional to the square of the divergence of their velocity field is approximately true. The relevant sort of question is whether or not the models that drive the simulations, which we build *using* artificial viscosity, accurately represent the real functional relationships that exist between the various properties of the fluid, at least approximately so. After all, it is not as if we use "the theory of artificial viscosity" by itself to make successful predictions and interventions. Its only when we couple this little bit of mathematical structure together with some form of the Navier-Stokes equations or the Euler equations—equations describing relationships between other variables of the fluid state—that we hope to make any useful predictions.

If we put the question this way, asking not whether the claim about artificial viscosity is approximately true, but asking rather whether the models we build—using artificial viscosity as one piece of the puzzle—are approximately realistic, the answer becomes less clear. Arguably, when each such model is considered individually and in its entirety, there is no compelling reason to deny that, while all of these models contain false features, some of them might very well count as reasonably accurately representing the relevant features of the fluid. There is arguably no compelling reason to resist viewing these models in quasi-realistic terms.

What I want to argue by way of reply to this objection is that defenders of the idea that success implies truth cannot simply avoid the first version of the question and hide behind the second. It is correct to say that it is the local models—models put together using fictions but also using many other model building principles—that are the engines of local successes. And it may indeed be arguable that these local models accurately represent, at least reasonably so. But it is also the case that the little bit of mathematical structure known as "artificial viscosity" is entitled to its own, in this case much less local, record of success. Indeed, it is only the artificial viscosity itself, and not the local models that incorporate it, that enjoys genuinely systematic and projectable success.

Recall a claim that I made above: Simulations are often used to learn about systems for which data are sparse. If such simulations are going to be at all useful, the models they use have to be trusted to produce good results despite all of the deviations from pure theory that go into making them. The credibility of those models comes not only from the credentials supplied to them by the governing theory, but also from the antecedently established credentials of the model-building techniques employed by the simulationist. Now take, for example, simulations of fluid flows with strong shocks. Dozens of different modeling schemes are used to model such systems. They use varying difference schemes, they exploit different symmetries, they deal with truncation error in different ways, and so on. One thing that the members of a large cross-section of these local models have in common is that they employ artificial viscosity to prevent unstable oscillations around the shocks. On a global basis, part of the reason we have for thinking that these local models are sanctionable is our conviction that artificial viscosity itself is a useful, off-the-shelf piece of mathematical structure that we can use successfully to build such models. In other words, it is not just the case that each of the models that employs artificial viscosity is itself locally capable of being used successfully to make predictions and interventions. It is also the case that artificial viscosity itself, when artfully and skillfully applied, can be successfully used for building many different sorts of local models in many different contexts. The success of artificial viscosity is far broader than that of any one of the local models that includes it. It is a piece of mathematical structure that has its own degree of success and trustworthiness in its own domain—a domain that is much larger than that of any of the local models that are built with its help.

Reliability without Truth

If we cannot say of artificial viscosity that it is true, or even approximately true, what property does this model-building principle have that allows us to expect it to yield sanctionable models—models that we trust in part *because* they include artificial viscosity in their set of equations? For those that believe that scientific success requires an explanation, what property of these model-building principles can we say accounts for our expectation of their future success? The right answer to this question, I think, is to say that model-building principles like artificial viscosity are *reliable*. Borrowing from an idea introduced by Arthur Fine, I mean to employ this term to designate a rival concept to truth—reliability.

In his attempt to deconstruct the realism/antirealism debate, Fine has argued that all arguments for realism based on the success of science fail because wherever the realist argues for truth, the instrumentalist can always settle for reliability, and vice versa (Fine 1991).

Instrumentalism takes reliability as its fundamental concept and differs from realism only in this: Where the realist goes for truth in the sense of a correspondence with reality, the instrumentalist goes for general reliability. . . . Where the realist says that science does (or should) aim at the truth, the instrumentalist says that science does (or should) aim at reliability. . . . The realist cannot win this game since whatever points to the truth, realist style, will also point to reliability.   (FINE 1996, 183)

I would note, of course, that the pragmatic notion of reliability that Fine is suggesting is quite distinct from the view often discussed in the epistemological literature known as "reliabilism." Reliabilism is a view about what further characteristics *true* beliefs need to have in order for us to count them as knowledge. On this view, the beliefs need to have been generated by a reliable process or method. In contrast, the notion of reliability being treated here is one that is meant to take the place of truth as a basic semantic notion. This contrast between the notion of reliabilism in epistemology and the notion of reliability being used here is clearly related to the discussion of technique versus claim discussed above. To take a *claim* to be reliable is to trust that we can rely upon it in much the same way that reliabilists demand that we be able to rely upon our justification-producing methods.

One other difference: reliabilists often define a reliable process or method in terms of something like the relative frequency with which it produces true results. Hence their notion of reliability is actually parasitic on truth. Clearly, our notion of reliability needs to avoid that. Hence, I characterize reliability (for modeling principles) in terms of being able to produce results that fit well into the web of our previously accepted data, our observations, the results of our paper-and-pencil analyses, and our physical intuitions, and to make successful predictions or produce engineering accomplishments.

Of course, the concept that Fine is talking about is *general reliability*. To take a claim as *generally* reliable "amount[s] to trusting it in all our practical and intellectual endeavors, . . . to be[ing] committed to understanding and dealing with the world from the perspective of that theory" (Fine 2001, 112).

Clearly, none of the contrary-to-fact model-building principles used in computer simulation can be taken to be *generally* reliable in the robust sense that Fine intends. Artificial viscosity has restricted scope—it takes art and skill to know how and when to apply it. It contradicts the content of our theory of fluids. For these and other reasons, we simply cannot be committed to understanding and dealing with the world from its perspective. But reliability, unlike truth, comes in degrees. Instrumentalists like Dewey, according to Fine, believed that "in inquiry we strive for concepts and theories that are generally reliable—although we often make do with less." If we take *"generally reliable"* to be a regulative ideal, then *"broadly reliable"* is a real-world instantiation.

This weaker notion of reliability dovetails well with much of the recent work in philosophy of science inspired by Nancy Cartwright's anti-fundamentalism—work that rejects the notion of universally true (or even generally reliable) theories and laws. In this tradition, theories and laws are seen as providing a *framework* for building models, schematizing experiments, and representing phenomena. They have very broad, but not universal, domains of application. Rather than taking theories and laws to be *universally true* and delimiting the character of all possible worlds, the anti-fundamentalist sympathizer takes them to be *broadly reliable* for a wide array of practical and epistemic tasks.

Fine remarked that many philosophers might find substituting reliability for truth to be nothing but a "semantic sleight of hand." Indeed, many fundamentalists feel the pull of the metaphysical intuition that behind any broadly reliable model-building principle must lie a universally true law. Successful model-building principles like artificial viscosity, however, provide a nice example to show that, at least in the case of *broad* reliability, no sleight of hand is involved. No semantic concept, not truth and not even approximate truth, adequately describes the proper attitude to have toward artificial viscosity and other model-building principles like it. The success of these models can thus provide a model of success in general: reliability without truth.

EIGHT

Conclusion

When computer simulation was pioneered as a scientific tool in the period directly following World War II, its use was limited to meteorology and nuclear weapons research. Since then, it has become indispensible in, and has even revolutionized, a growing number of disciplines. The list of sciences that make extensive use of computer simulation has grown to include astrophysics, materials science, engineering, fluid mechanics, climate science, evolutionary biology, ecology, economics, decision theory, sociology, and many others. This much is clear: In terms of its centrality to a sheer quantity and variety of innovation, the last several decades of the history of science have been the age of computer simulation.

Normally, such deep and widespread changes in the way science is practiced—in the way knowledge, in such a wide variety of domains, is acquired—would attract the immediate interest of philosophers of science. This did not happen. It has not happened, I have argued, because philosophers of science have had a bias in favor of the proposition that the philosophically interesting action in the sciences occurs when new theories are proposed. Philosophers have always been interested in revolutionary changes, but the presumption has been that the interesting changes would always come in the form of new fundamental descriptions of the world.

But in the introduction, I urged readers to consider the possibility that it was not only changes in basic theory that could be of interest to philosophers—to consider the possibility that new experimental methods, new research technologies, or innovative ways of solving new sets of problems within existing theory could have a similar impact on philosophy. And I suggested that philosophers might find as good philosophical fodder by examining the details of how scientists model complex phenomena within existing theories as they do by looking to novel fundamental theories.

It is time to revisit and evaluate those claims. I think they hold up well. A close look at the use of computer simulation has taught us to appreciate the importance of a much more nuanced view of the relation between theory and its applications. Theory can stand in a wide variety of relations to its applications; sometimes theory is applied directly—in a process that is well captured by the idea of derivation—and sometimes the path from theory to application is much more indirect, with theory playing only a contributing role in generating local representations of phenomena. There is, consequently, a whole category of epistemological issues in the sciences that have escaped the attention of philosophers, who have traditionally concerned themselves with the justification of theories and not with their application.

We have, in other words, rejected the overly conservative intuition that computer simulation is nothing but boring and straightforward theory application. But we have avoided embracing the opposite, overly grandiose intuition: that simulation is a radically new kind of knowledge production, "on a par" with experimentation. In fact, we have seen that soberly locating simulation "on the methodological map" is not a simple matter.

On the one hand, there is much to be gained by drawing comparisons between simulation and experiment. Much of what we have learned about the epistemology of simulation has drawn heavy inspiration not only from recent work on the autonomy of models, but also directly from the philosophy of experiment. Drawing on the work of Alan Franklin, who identified a number of strategies that experimenters use to increase rational confidence in their results, we saw how simulationists use many analogous strategies to do the same for their own results. And drawing on the work of Ian Hacking, who argued for the claim that "experiments have a life of their own," we saw how arguments for the reliability of many simulation results depend, similarly, on computational methods having their own life as well. The techniques that simulationists use to construct their models get credentialed in much the same way that Hacking says that instruments and experimental procedures and methods do: the credentials develop over an extended period of time and become deeply tradition-bound. In Hacking's language, the techniques and sets of assumptions that simulationists use become "self-vindicating." The credibility, in other words, of discretization and finite-element methods, solvers, parameterization schemes, multiscale modeling, handshaking algorithms, modeling fictions, and so on cannot be argued for entirely on mathematical or theoretical grounds—they bring their own reliability, established in their own disciplinary traditions.

On the other hand, what we might call the ontological relationship between simulations and experiments is quite complicated. Is it true that simulations are, after all, a particular species of experiment? I have tried to argue against this claim, while at the same time insisting that the differences between simulation and experiment are more subtle than some of the critics of the claim have suggested. Most important, I have tried to argue that we should disconnect questions about the identity of simulations and experiments from questions of the epistemic power of simulations. Despite (or sometimes even in virtue of) having certain characteristic features that are distinct from those of experiments, simulations sometimes have great epistemic power—there are indeed some questions we can ask for which simulations provide much more reliable answers than any experiment can provide. At the same time, thinking carefully about the relationship between experiments and simulations can teach a great deal, I would argue, about experiment itself, and particularly about the role played by models and background knowledge.

We have also seen how some specific kinds of simulations raise special philosophical issues. One of these is surely the so-called parallel multiscale simulations we saw in chapter 5 that draw on theories from more than one level of description. Such simulations put pressure on two philosophical intuitions that are often taken for granted. The first intuition is that an inconsistent set of laws can have no models. Strictly speaking, this is of course true. But it is often assumed that the model of axiomatic logic and semantics is a sufficiently good rational reconstruction of theory application for it to follow from this that an inconsistent set of theoretical principles can give rise to no models. But sometimes simulationists do build models "out of" an inconsistent set of theoretical claims. The second intuition is that the interesting relationships between theories at different levels of description are fully captured by the degree to which the higher-level theory is reducible to the lower-level theory in the ordinarily understood way. In fact, these same examples seem to show that it is a delicate and empirical matter how different theories will relate to each other in a successful and reliable model. Another special kind of simulation that raises its own kind of philosophical issues is the class of simulations used—particularly in guiding policy—to predict the future of the earth's climate. The high degree of uncertainty associated with the predictions of these models and their significant policy implications combine to make an interesting background for examining the role of social values in the appraisal of scientific hypotheses.

Simulations used to predict the future of the earth's climate also give rise to their own special philosophical issues. Two things make these special. The first is their complexity and degree of historical entrenchment, and the second is the combination of the degree of uncertainty associated with them and their significant public policy implications. As a result of these features, consideration of these models adds complexity to the debate about the appropriate role of social, political, and ethical values in the appraisal of scientific hypotheses.

And finally, we saw how certain kinds of modeling assumptions used in simulations put pressure on one of the central arguments in favor of scientific realism: arguments based on the idea that success implies truth. We saw, in particular, that many principles employed in simulations can be highly reliable without being even approximately true.

The philosophy of science should continue, as it always has in the past, to respond to the character of the science of its own era. If much of contemporary science has indeed entered *the age of computer simulation*, then philosophy of science should respond accordingly. And surely much work remains to be done. One of the interesting things about advances in the computationally intensive sciences is that they are often advances in overcoming present practical problems. We use the multiscale methods described in chapter 5, for example, because it is presently impossible, in practice, to model micron-sized pieces of material entirely at the quantum mechanical level. But what is practically impossible today may become easy to accomplish tomorrow. We will surely someday have computers fast enough and powerful enough to solve, by brute force—and in a more principled way—problems that today we solve using the kinds of clever tricks that have inspired many of the philosophical ideas put forth in this book. When that happens, will we, in the course of continually pushing the envelope of problems we wish to solve, employ new tricks that support similar philosophical conclusions? Or will we eventually bring all the phenomena that interest us under the umbrella of principled theoretical application? "I shall take care not to risk a prophecy which might be falsified between the day this book is ready for the press and the day on which it is placed before the public."[1]

---

1. Poincaré 1952, 244.

References

Abraham, F., Jeremy Q. Broughton, Noam Bernstein, and Efthi-
    mios Kaxiras. 1998. "Spanning the length scales in dynamic
    simulation." *Computers in Physics* 12 (6): 538–46.
Allen, Myles 2008. "What can be said about future climate?"
    Available online at http://www.climateprediction.net/
    science/pubs/allen_Harvard2008.ppt (accessed July 3, 2008).
American Institute of Aeronautics and Astronautics. 1998. *AIAA
    Guide for the Verification and Validation of Computational Fluid
    Dynamics Simulations*. AIAA G-077–1998. Reston, VA: Ameri-
    can Institute of Aeronautics and Astronautics.
Batterman, R. 2002. *The Devil in the Details: Asymptotic Reasoning
    in Explanation, Reduction, and Emergence*. New York: Oxford
    University Press.
Benestad, R. 2007 (May 27). "Why global climate models do not
    give a realistic description of the local climate." *RealClimate*.
    Available online at www.realclimate.org (retrieved May 29,
    2007).
Broughton, J., Farid F. Abraham, Noam Bernstein, and Efthimios
    Kaxiras. 1999. "Concurrent coupling of length scales: Meth-
    odology and application." *Physical Review B* 60 (4): 2391–403.
Campbell, D. 1957. "Factors relevant to the validity of experi-
    ments in social settings." *Psychological Bulletin* 54:297–312.
Campbell, J. 2000. "Artificial viscosity for multi-dimensional hy-
    drodynamics codes." Available online at http://cnls.lanl
    .gov/Highlights/2000–09/article.htm.
Caramana, E. J., M. J. Shashkov, and P. P. Whalen. 1988. "Formu-
    lations of artificial viscosity for multi-dimensional shock
    wave computations." *Journal of Computational Physics* 144:
    70–97.
Cartwright, N. 1983. *How the Laws of Physics Lie*. Oxford: Oxford
    University Press.

———. 1989. *Nature's Capacities and Their Measurement*. Oxford: Oxford University Press.

———. 1999. *The Dappled World, a Study of the Boundaries of Science*. Cambridge: Cambridge University Press.

Chalmers, David. 1996. "Does a rock implement every finite-state automaton?" *Synthese* 108:309–33.

Churchman, C. West. 1948. *Theory of Experimental Inference*. New York: Macmillan.

———. 1956. "Science and Decision Making." *Philosophy of Science* 22:247–49.

Clark, Andy. 1987. "The kluge in the machine." *Mind and Language* 2:277–300.

Committee on the Evaluation of Quantification of Margins and Uncertainties Methodology for Assessing and Certifying the Reliability of the Nuclear Stockpile, National Research Council. 2008. *Evaluation of Quantification of Margins and Uncertainties Methodology for Assessing and Certifying the Reliability of the Nuclear Stockpile*. Washington, DC: National Academies Press.

Darling, Karen Merikangas. 2002. "The complete Duhemian underdetermination argument: Scientific language and practice." *Studies in History and Philosophy of Science* 33A:511–33.

Dewey, John. 1929. *The Quest for Certainty*. New York: Minton, Balch. Reprinted in Jo Ann Boydston (ed.), *John Dewey: The Later Works, 1925–1953*. Vol. 4. Carbondale: Southern Illinois University Press, 1988.

Douglas, Heather. 2000. "Inductive risk and values in science." *Philosophy of Science* 67:559–79.

———. 2004a. "Border skirmishes between science and policy: Autonomy, responsibility, and values." In Machamer and Wolters 2004, 220–44.

———. 2004b. "The irreducible complexity of objectivity." *Synthese* 138:453–73.

Dowling, D. 1999. "Experimenting on Theories." *Science in Context* 12 (2): 261–73.

Elgin, C. 2009. "Exemplification, idealization, and scientific understanding." In Suarez 2009, 77–90.

Fine, A. 1991. "Piecemeal realism." *Philosophical Studies* 61:79–96.

———. 1993. "Fictionalism." *Midwest Studies in Philosophy* 18:1–18.

———. 1996. *The Shaky Game*. Chicago: University of Chicago Press.

———. 1998. "The viewpoint of no-one in particular." *Proceedings and Addresses of the American Philosophical Association* 72:9–20.

———. 2001. "The scientific image twenty years later." *Philosophical Studies* 106:107–22.

Frank, Philip G. 1954. "The variety of reasons for the acceptance of scientific theories." In *The Validation of Scientific Theories*, ed. Philip Frank, 3–17. Boston: Beacon Press.

Franklin, A. 1986. *The Neglect of Experiment*. New York: Cambridge University Press.

Frigg, Roman, and Julian Reiss. 2009. "The philosophy of simulation: Hot new issues or same old stew." *Synthese* 169:593–613.

Frisch, M. 2004. "Inconsistency in classical electrodynamics." *Philosophy of Science* 71 (4): 525–49.

Galison, P. 1996. "Computer Simulations and the Trading Zone." In *The Disunity of Science: Boundaries, Contexts, and Power*, ed. P. Galison and D. Stump, 118–57. Stanford, CA: Stanford University Press.

———. 1997. *Image and Logic: A Material Culture of Microphysics*. Chicago: University of Chicago Press.

Giere, R. 1988. *Constructing Science*. Chicago: University of Chicago Press.

———. 1999. *Science without Laws*. Chicago: University of Chicago Press.

———. 2003. "A new program for philosophy of science?" *Philosophy of Science* 70:15–21.

Gilbert, N., and K. Troitzsch. 1999. *Simulation for the Social Scientist*. Philadelphia, PA: Open University Press.

Gleckler, P. J., K. E. Taylor, and C. Doutriaux. 2008. "Performance metrics for climate models." *Journal of Geophysical Research —Atmospheres* 113 (6): 104–21.

Goldstein, M., and J. C. Rougier. 2009, "Reified Bayesian modelling and inference for physical systems." *Journal of Statistical Planning and Inference* 139 (3): 1221–39.

Guala, F. 2002. "Models, simulations, and experiments." In *Model-Based Reasoning: Science, Technology, Values*, ed. Lorenzo Magnani and Nancy Nersessian, 59–74. New York: Kluwer.

Hacking, I. 1983. *Representing and Intervening: Introductory Topics in the Philosophy of Science*. New York: Free Press.

———. 1988. "On the stability of the laboratory sciences." *Journal of Philosophy* 85 (10): 507–15.

———. 1992. "Do thought experiments have a life of their own?" In *Proceedings of the 1992 Biennial Meeting of the Philosophy of Science Association*, ed. A. Fine, M. Forbes, and K. Okruhlik, 2:302–10. East Lansing, MI: The Philosophy of Science Association.

Hempel, Carl G. 1945. "Studies in the logic of confirmation." *Mind* 54:1–26. Reprinted in Hempel 1965, 3–46, with a 1964 postscript added.

———. 1965. *Aspects of Scientific Explanation and Other Essays in the Philosophy of Science*. New York: Free Press.

Hempel, Carl G., and Paul Oppenheim. 1948. "Studies in the logic of explanation." *Philosophy of Science* 15:135–75. Reprinted in Hempel 1965, 245–90, with a 1964 postscript added.

Horwich, P. 1999. *Truth*. Oxford: Oxford University Press.

Howard, Don A. 2006. "Lost wanderers in the forest of knowledge: Some thoughts on the discovery-justification distinction." In *Revisiting Discovery and Justification: Historical and Philosophical Perspectives on the Context Distinction*, ed. J. Schickore and F. Steinle, 3–22. New York: Springer.

Hughes, R. 1999. "The Ising model, computer simulation, and universal physics." In Morgan and Morrison 1999, 97–145.

Humphreys, P. 1991. "Computer simulation." In *Proceedings of the 1990 Biennial Meeting of the Philosophy of Science Association*, ed. A. Fine, M. Forbes, and L. Wessels, 2:497–506. East Lansing, MI: The Philosophy of Science Association.

———. 1994. "Numerical experimentation." In *Philosophy of Physics, Theory Structure, Measurement Theory, Philosophy of Language, and Logic*. Vol. 2 of *Patrick Suppes: Scientific Philosopher*, ed. P. Humphreys, 103–24. Dordrecht: Kluwer.

———. 1995. "Computational science and scientific method" in *Minds and Machines* 5 (1):499–512.

IPCC. 2007. *Climate Change 2007—The Physical Science Basis. Contribution of Working Group I to the Fourth Assessment Report of the Intergovernmental Panel on Climate Change*. Cambridge: Cambridge University Press.

Jacobs, M. L., D. H. Porter, and P. R. Woodward. 1998. "3-D numerical models of red giant stars." Abstract no.193. American Astronomical Society Meeting.

Jeffrey, Richard C. 1956. "Valuation and acceptance of scientific hypotheses." *Philosophy of Science* 22:237–46.

Kaufmann, W. J., and L. L. Smarr. 1993. *Supercomputing and the Transformation of Science*. New York: Scientific American Library.

Kitcher, Philip. 1993. *The Advancement of Science: Science without Legend, Objectivity without Illusions*. New York: Oxford University Press.

———. 2001. *Science, Truth, and Democracy*. New York: Oxford University Press.

———. 2002. "On the explanatory role of correspondence truth." *Philosophy and Phenomenological Research* 64:346–64.

Koertge, Noretta. 2003. "Screen out social values from core epistemic areas of science." In *Scrutinizing Feminist Epistemology: An Examination of Gender in Science*, ed. C. Pinnick, R. Almeder, and N. Koertge, 47–64. New Brunswick, NJ: Rutgers University Press.

Kourany, Janet. 2003a. "A philosophy of science for the twenty-first century." *Philosophy of Science* 70:1–14.

———. 2003b. "Reply to Giere." *Philosophy of Science* 70:22–26.

Kuhn, Thomas 1962. *The Structure of Scientific Revolutions*. Chicago: University of Chicago Press.

———. 1969. "Postscript–1969." In *The Structure of Scientific Revolutions*, 174–210. 2nd ed. Chicago: University of Chicago Press.

———. 1977. "Objectivity, value judgment, and theory choice." In *The Essential Tension: Selected Studies in Scientific Tradition and Change*, by Thomas Kuhn, 320–39. Chicago: University of Chicago Press.

———. 1977. "The function of measurement in modern physical science." In *The Essential Tension: Selected Studies in Scientific Tradition and Change*, by Thomas Kuhn, 161–93. Chicago: University of Chicago Press.

Laudan, L. 1981. "A confutation of convergent realism." *Philosophy of Science* 48:218–49.

Laymon, R. 1985. "Idealizations and the testing of theories by experimentation." In *Observation, Experiment and Hypothesis in Modern Physical Science*, ed. P. Achinstein and O. Hannaway, 147–73. Cambridge, MA: MIT Press.

———. 1991. "Computer simulations, idealizations and approximations." In *Proceedings of the 1990 Biennial Meeting of the Philosophy of Science Association*, ed. A. Fine, M. Forbes, and L. Wessels, 2:519–34. East Lansing, MI: The Philosophy of Science Association.

Lenhard, Johannes. 2007. "Computer simulation: The cooperation between experimenting and modeling." *Philosophy of Science* 74:176–94.

Longino, Helen. 1990. *Science as Social Knowledge: Values and Objectivity in Scientific Inquiry*. Princeton, NJ: Princeton University Press.

———. 1996. "Cognitive and non-cognitive values in science: Rethinking the dichotomy." In *Feminism, Science, and the Philosophy of Science*, ed. L. H. Nelson and J. Nelson, 39–58. Dordrecht: Kluwer.

———. 2002. *The Fate of Knowledge*. Princeton, NJ: Princeton University Press.

Machamer, Peter, and Gereon Wolters, eds. 2004. *Science, Values, and Objectivity*. Pittsburgh: University of Pittsburgh Press.

Marion, M. 1998. *Wittgenstein, Finitism, and the Foundations of Mathematics*. Oxford: Clarendon Press.

McMullin, Ernan. 1983. "Values in science." In *Proceedings of the 1982 Biennial Meeting of the Philosophy of Science Association*, ed. P. D. Asquith and T. Nickles, 1:3–28. East Lansing, MI: Philosophy of Science Association.

Mitchell, Sandra D. 2004. "The prescribed and proscribed values in science policy." In Machamer and Wolters 2004, 245–55.

Moin, P., and J. Kim. 1997. "Tackling turbulence with supercomputers." *Scientific American* 276 (1): 62–68.

Morgan, M. 2002. "Model experiments and models in experiments." In *Model-Based Reasoning: Science, Technology, Values*, ed. Lorenzo Magnani and Nancy Nersessian, 41–58. New York: Kluwer.

———. 2003. "Experiments without material intervention: Model experiments, virtual experiments and virtually experiments." In *The Philosophy of Scientific Experimentation*, ed. Hans Radder, 216–35. Pittsburgh, PA: University of Pittsburgh Press.

Morgan, M., and M. Morrison, eds. 1999. *Models as Mediators*. Cambridge: Cambridge University Press.

Morrison, M. C. 1998. "Modeling nature: Between physics and the physical world." *Philosophia Naturalis* 54:65–85.

Nakano, Aiichiro, Martina E. Bachlechner, Rajiv K. Kalia, Elefterios Lidorikis, Priya Vashishta, George Z. Voyiadjis, Timothy J. Campbell, Shuji Ogata, and Fuyuki Shimojo. 2001. "Multiscale simulation of nano-systems." *Computing in Science and Engineering* 3 (4): 56–66.

Neurath, Otto. 1913. "Die Verirrten des Cartesius und das Auxiliarmotiv. Zur Psychologie des Entschlusses." In *Jahrbuch der Philosophischen Gesellschaft an der Universität Wien*. Leipzig: Johann Ambrosius Barth. Published in English as "The lost wanderers of Descartes and the auxiliary motive (on the psychology of decision)." In *Philosophical Papers, 1913–1946*, ed. and trans. R. S. Cohen and M. Neurath, 1–12. Dordrecht: D. Reidel.

Norton, S., and Suppe, F. 2001. "Why atmospheric modeling is good science." In *Changing the Atmosphere: Expert Knowledge and Environmental Governance*, ed. C. Miller and P. Edwards, 88–133. Cambridge, MA: MIT Press.

Parker, W. 2009. "Does matter really matter: Computer simulations, experiments, and materiality." *Synthese* 169:483–96.

Poincaré, H. 1952. *Science and Hypothesis*. New York: Dover.

Porter, D., and P. Woodward. 1994. "High-resolution simulations of compressible convection using the piecewise-parabolic method." *Astrophysical Journal* 93 (supp.): 309–21.

Porter, D., S. Anderson, and P. Woodward. 1998. "Simulating a pulsating red giant star." Preprint (no info. available).

Rohrlich, F. 1991. "Computer simulation in the physical sciences." In *Proceedings of the 1990 Biennial Meeting of the Philosophy of Science Association*, ed. A. Fine, M. Forbes, and L. Wessels, 2:507–18. East Lansing, MI: The Philosophy of Science Association.

Roy, Subrata. 2005. "Recent advances in numerical methods for fluid dynamics and heat transfer" (editorial). *Journal of Fluid Engineering* 127 (4): 629–30.

Rudd, R. E., and J. Q. Broughton. 2000. "Concurrent coupling of length scales in solid state systems." *Physica Status Solidi B* 217:251–91.

Rudner, Richard. 1953. "The scientist *qua* scientist makes value judgments." *Philosophy of Science* 20:1–6.

Shirayama, S., and K. Kuwahara. 1990. "Flow visualization in computational fluid dynamics." *International Journal of Supercomputer Applications* 4 (2): 66–80.

Simon, Herbert. 1969. *The Sciences of the Artificial*. Boston, MA: MIT Press.

Siono, Sigekata, ed. 1962. *Proceedings of the International Symposium on Numerical Weather Prediction* (Tokyo, November 1960). Tokyo: Meteorological Society of Japan.

Smarr, L. 1985. "An approach to complexity" *Science* 228:403–8.

Smith, Leonard A. 2002. "What might we learn from climate forecasts?" *Proceedings of the National Academy of Sciences USA* 4 (99): 2487–92.

Solomon, Miriam. 2001. *Social Empiricism*. Cambridge: MIT Press.

Steinhoff, J., and Underhill, D. 1994. "Modification of the Euler equations for 'vorticity confinement': Application to the computation of interacting vortex rings." *Physics of Fluids* 6:2738–44.

Stern, Nicholas. 2007. *The Economics of Climate Change: The Stern Review*. Cambridge: Cambridge University Press.

Stillinger, F. H., and T. A. Weber. 1985. "Computer simulation of local order in condensed phases of silicon." *Physical Review B* 31:5262–71.

Suarez, M., ed. 2009. *Fictions in Science: Philosophical Essays on Modeling and Idealization*, London: Routledge.

Suppe, F., ed. 1974. *The Structure of Scientific Theories*. Urbana: University of Illinois Press.

———. 1989. *The Semantic Conception of Theories and Scientific Realism*. Urbana: University of Illinois Press.

Suppes, P. 1962. "Models of Data." In *Logic, Methodology and Philosophy of Science*, ed. E. Nagel, P. Suppes, and A. Tarski, 252–61. Stanford, CA: Stanford University Press.

Unruh, W. G. 1981. "Experimental black-hole evaporation?" *Physical Review Letters* 46 (21): 1351–53.

———. 2002. "Measurability of dumb hole radiation?" In *Artificial Black Holes*, ed. M. Novello, M. Visser, and G. E. Volovik, 199–122. New Jersey: World Scientific.

Vaihinger, Hans. *The Philosophy of 'As If': A System of the Theoretical, Practical and Religious Fictions of Mankind*. Trans. C. K. Ogden. New York: Barnes and Noble, 1968. (Orig. pub. in England by Routledge and Kegan Paul, 1924.)

Van Fraassen, B. C. 1970. "On the extension of Beth's semantics of physical theories." *Philosophy of Science* 37:325–38.

———. 1980. *The Scientific Image*. Oxford: Oxford University Press.

von Neumann J., and R. D. Richtmyer. 1950. "A method for the numerical calculation of hydrodynamical shocks." *Journal of Applied Physics* 21:232–47.

Visser, M. 2002. "Introduction and survey." In *Artificial Black Holes*, ed. M. Novello, M. Visser, and G. E. Volovik, 1–35. New Jersey: World Scientific.

Weissert, T. 1997. *The Genesis of Simulation in Dynamics: Pursuing the Fermi-Pasta-Ulam Problem*. New York: Springer.

Wilhelmson, R. B., Brian F. Jewett, Crystal Shaw, Louis J. Wicker, Matthew Arrott, Colleen B. Bushell, Mark Bajuk, Jeffrey Thingvold, and Jeffery B. Yost. 1990. "A study of the evolution of a numerically modeled severe storm." *International Journal of Supercomputer Applications* 4 (2): 20–36.

Wilholt, Torsten. 2009. "Bias and values in scientific research." *Studies in History and Philosophy of Science* 40 (1): 92–101.

Wimsatt, William. 2007. *Re-engineering Philosophy for Limited Beings: Piecewise Approximations to Reality*. Cambridge, MA: Harvard University Press.

Winkler, K., J. Chalmers, S. Hodson, P. Woodward, and N. Zabusky. 1987. "A numerical laboratory." *Physics Today* 40 (10): 28–37.

Winsberg, E. 1999a. "Sanctioning models: The epistemology of simulation." *Science in Context* 12 (2): 275–92.

———. 1999b. " Simulation and the philosophy of science: Computationally intensive studies of complex physical systems." PhD diss., Indiana University.

———. 2001. "Simulations, models and theories: Complex physical systems and their representations." *Philosophy of Science* 68 (supp.): S442–58.

———. 2003. "Simulated experiments: Methodology for a virtual world." *Philosophy of Science* 70:105–25.

———. 2006. "Models of success versus the success of models: Reliability without truth." *Synthese* 152 (1): 1–19.

———. 2006. "Handshaking your way to the top: Simulation at the nanoscale." *Philosophy of Science* 73:582–94.

———. Forthcoming. "A Function for Fictions."

Zabusky, N. 1987. "Grappling with complexity." *Physics Today* 40 (10): 25–27.

Index

Milton Keynes UK
Ingram Content Group UK Ltd.
UKHW011828300923
429712UK00002B/67

9 780226 902029